Accessing and Sharing the Benefits of the Genomics Revolution

The International Library of Environmental, Agricultural and Food Ethics

VOLUME 11

Editors

Michiel Korthals, *Dept. of Applied Philosophy, Wageningen University, Wageningen, The Netherlands*
Paul B. Thompson, *Dept. of Philosophy, Michigan State University, East Lansing, U.S.A.*

Editorial Board

Andrew Brennan, *The University of Western Australia, Australia*
Avner de-Shalit, *Hebrew University, Jerusalem, Israel*
Clare Palmer, *Washington University, St Louis, U.S.A.*
Daryl Macer, *The Eubios Ethics Institute, University of Tsukuba, Ibaraki, Japan*
Doris Schroder, *University of Central Lancashire, Preston, United Kingdom*
Lawrence Busch, *Dept. of Sociology, Michigan State University, Lansing, U.S.A.*
Richard Haynes, *Dept. of Philosophy, University of Florida, Gainesville, U.S.A.*

ACCESSING AND SHARING THE BENEFITS OF THE GENOMICS REVOLUTION

Edited by

Peter W.B. Phillips
University of Saskatchewan, Saskatoon, SK, Canada

and

Chika B. Onwuekwe
University of Calgary, AB, Canada

A C.I.P. Catalogue record for this book is available from the Library of Congress.

ISBN 978-1-4020-5821-9 (HB)
ISBN 978-1-4020-5822-6 (e-book)

Published by Springer,
P.O. Box 17, 3300 AA Dordrecht, The Netherlands.

www.springer.com

Printed on acid-free paper

All Rights Reserved
© 2007 Springer
No part of this work may be reproduced, stored in a retrieval system, or transmitted in
any form or by any means, electronic, mechanical, photocopying, microfilming,
recording or otherwise, without written permission from the Publisher, with the exception
of any material supplied specifically for the purpose of being entered and executed on a
computer system, for exclusive use by the purchaser of the work.

CONTENTS

Contributors vii

Acknowledgements ix

Part One: ACCESS AND BENEFITS SHARING IN CONTEXT

1. Introduction to the Challenge of Access and Benefit Sharing 3
   Peter W.B. Phillips and Chika B. Onwuekwe

Part Two: SHARING THE BENEFITS OF INVENTIONS, PGRS AND TRADITIONAL KNOWLEDGE

2. Ideology of the Commons and Property Rights: Who Owns Plant Genetic Resources and the Associated Traditional Knowledge? 21
   Chika B. Onwuekwe

3. Farmers' Privilege and Patented Seeds 49
   Peter W.B. Phillips

4. Traditional Knowledge and Benefit Sharing: From Compensation to Transaction 65
   David Castle and E. Richard Gold

5. Biological Resources, Intellectual Property Rights and International Human Rights: Impacts on Indigenous and Local Communities 81
   Donna Craig

6. Lost in Translation? The Rhetoric of Protecting Indigenous Peoples' Knowledge in International Law and the Omnipresent Reality of Biopiracy 111
   Ikechi Mgbeoji

Part Three: IMPLEMENTING ACCESS AND BENEFITS SHARING

7. Liability Principles and their Impact on Access and Benefits Sharing 145
   Lara Khoury

8. Beyond the Rhetoric: Population Genetics and Benefit-Sharing 157
   Lorraine Sheremeta and Bartha Maria Knoppers

9. Bioprocessing Partnerships in Practice: A Decade of Experiences at INBio in Costa Rica 183
   Jorgé Cabrera Medaglia

Part Four: ACCESS AND BENEFIT SHARING IN THE NEW MILLENNIUM

10. Conclusions: New Paths to Access and Benefit Sharing 199
    Chika B. Onwuekwe and Peter W.B. Phillips

Index 209

CONTRIBUTORS

Dr. David Castle holds the Canada Research Chair in Science and Society in the department of Philosophy at the University of Ottawa, Canada. His research and teaching interests lie in the philosophy of the life sciences, with particular emphasis on evolutionary biology and ecology, environmental philosophy and the ethical implications posed by biotechnology. He is an investigator or principal investigator on a variety of large-scale applied ethics research programs related to biotechnology.

Donna Craig is Professor in Law and the Co-Director of the Environmental Law Centre at Macquarie University, and Professor of Desert Knowledge, Institute of Advanced Studies, Charles Darwin University. She is also a specialist practitioner in the area of international and national environmental law and policy. Donna has over 25 years experience in research, legal practice, teaching and working with communities, indigenous peoples' organizations, governments and corporations.

Dr. Richard Gold is the Director of the Centre for Intellectual Property Policy and teaches in the area of intellectual property and common law property at McGill University's Faculty of Law. His research centres on the nexus between technology, commerce, and ethics, particularly with respect to biotechnology in the international context. He is the Principal Investigator of the Intellectual Property Modeling Group, a transdisciplinary research team investigating intellectual property regimes.

Dr. Lara Khoury is an Assistant Professor at the Faculty of Law at McGill University in Montreal, Canada, where she teaches and conducts research from comparative and trans-systemic perspectives, particularly in the fields of Medical and Environmental Liability. She holds D.Phil. and B.C.L. degrees from the University of Oxford (U.K.).

Dr. Bartha Maria Knoppers is Canada Research Chair in Law and Medicine, and Professor at the Faculté de droit, Université de Montréal, Canada, Senior Researcher at the Centre for Public Law (C.R.D.P.), holder of the Chaire d'excellence Pierre Fermat (France) and Officer of the Order of Canada and Fellow of the AAAS.

Professor Jorge Cabrera Medaglia is Lead Counsel for International Sustainable Biodiversity Law with the CISDL, a Professor of Environmental Law at University of Costa Rica, and a Professor of International Trade for the School of Business Administration, University of Costa Rica. He is also a member of the UNEP initiative on capacity building on access to genetic resources and the National Biodiversity Commission of Costa Rica. He has acted as a negotiator at the Convention on Biological Diversity (CBD) on behalf of the Government of Costa Rica, served as co-chair of the CBD's Expert Panel on Access and Benefit Sharing and chair of the CBD Sub-working Group on IPR and Capacity-Building.

CONTRIBUTORS

Dr. Ikechi Mgbeoji is Associate Professor at Osgoode Hall Law School of York University, Toronto, Canada. Educated in Nigeria, Canada, and Germany, he practiced for five years in civil litigation specializing in Commercial Litigation and Intellectual Property Law before reentering academia. His main research and teaching interests are in the areas of international comparative and transnational law; international intellectual property law; and environmental law.

Dr. Chika B. Onwuekwe is an attorney-at-law (Canada and Nigeria) and Professor of Law and Society at the University of Calgary, Alberta, Canada. Dr. Onwuekwe's research interests are in the legal, economic, governance and social issues of transformative technologies, emerging democracies, energy and natural resources law, and policy issues around intellectual property and resource allocation. He practices, teaches and publishes in these and related areas.

Dr. Peter W.B. Phillips, an international political economist, is Professor of Political Studies and an associate member of the departments of Agricultural Economics and Management at the University of Saskatchewan, Canada and holds a concurrent faculty appointment as Professor at Large at the Institute for Advanced Studies, University of Western Australia, Perth. His research concentrates on issues related to governing transformative innovations. He is the co-principal investigator of the Genome Alberta project on Translating Knowledge in Health Systems (2006-2010) and leader of the sub-project investigating public-private partnerships related to access and benefits sharing of plant genetics resources.

Lorraine Sheremeta is a lawyer and Research Associate at the Health Law Institute, Faculty of Law, University of Alberta, Canada. Her work focuses primarily on emerging technologies (including genomics, stem cell technologies and nanotechnology) commercialization, and law.

ACKNOWLEDGEMENTS

This book is a product of the Genome Prairie Intellectual Property Rights Conference held in Banff, Canada, on January 29 - February 1, 2004. We invited 25 knowledgeable individuals, including academics, stakeholders and practitioners from North America and around the world, to debate the nature of the current intellectual property system and its influence on traditional knowledge and access to benefits sharing related to the products of the genomics revolution.

In addition to the participants and authors, we would like to acknowledge Genome Prairie for the financial support to run the conference. We also must acknowledge the diligent efforts of Alexis Dahl, Julie Graham, Jen Medlock and Tara Procyshyn to get this work into this form.

Peter W.B. Phillips, Saskatoon
Chika B. Onwuekwe, Calgary

PART ONE

ACCESS AND BENEFITS SHARING IN CONTEXT

PETER W.B. PHILLIPS AND CHIKA B. ONWUEKWE

CHAPTER 1

INTRODUCTION TO THE CHALLENGE OF ACCESS AND BENEFIT SHARING

For decades, there has been an ongoing and intense debate about the appropriateness of patents for living matter. This debate was greatly intensified in the late 1990s with the convergence of three developments. First, advances in science led to a rapid increase in the number of biotechnology patents in the late 1990s, particularly in the field of genetics. Second, the international community negotiated in 1995 a new international regime for intellectual property protection through the World Trade Organization's Agreement on Trade-Related Aspects of Intellectual Property Rights (TRIPs Agreement). Third, international debate about ownership, control, access and benefits sharing related to global plant genetic resources heated up through a variety of negotiating venues under the auspices of the Convention on Biological Diversity (CBD). These three developments at times supported and at other times challenged the international innovation agenda that is increasingly focused on knowledge-based economic growth.

The expansion of private rights to intellectual property, particularly for inventions related to composition of living matter and genetic isolates, is viewed as the foundation of knowledge-based economic growth. This has generated significant public debate. The dominant agenda generally operates on the assumption that patents are essential to the identification of research targets, the mobilization of public and private resources, the oversight of new technologies and products and the active adaptation, adoption and use of new end-products. Those supporting this extension argue that, in a period of knowledge-based growth, private investors need to have the incentive for and assurances of monopoly access to and use of their inventions. Many of these supporters firmly believe that private initiative is the only way to generate the optimal amount of investment in Research and Development (R&D) and that any resulting inequities should be handled outside the formal Intellectual Property (IP) system. Arrays of opponents challenge this expansion of private rights, arguing either that the range and scope of rights may be excessive or, in some cases, inappropriate.

This issue is important for everyone involved in agri-food research and general development policy. The debate and related conflicts about access and benefits sharing do not align exclusively along a North–South axis. While many of the higher profile conflicts relate to germplasm drawn from indigenous communities in "southern" developed countries, there is rising interest and concern about how indigenous communities in Canada, the US, Australia and many other developed countries will be able to control access and share benefits related to use of their traditional knowledge and genetic resources. Similarly, virtually all agri-food research programs in all nations depend heavily on germplasm from others—no country is self-sufficient in any major crop. Hence, how the debate unfolds and how rights are sustained or realigned will have an influence around the world.

The chapters in this book offer a range of frameworks for analyzing the issues and assumptions that underlie the varying perspectives on the appropriate use of patents on higher life forms. Aside from the moral concerns associated with "life patents", an area we discuss but expect is not amenable to consensus, we ask a variety of questions that should provide a basis for consensus building in this contentious policy area. Specifically, is there really any profound social or economic problem with patents on higher life forms? That is, do gene patents create more problems than benefits? If so, what is the nature and scope of the problems? Are these problems likely to be enduring, such that reform to the patent system is needed? Or are the problems related to the relative novelty of patents in this area? Can changes be made that will make matters better? And, given that much of the debate is about access to and sharing the benefits of traditional, contemporary and anticipated improvements in living organisms, we also must ask whether we need to substantively address access and benefits sharing and, if so, should that be a common responsibility?

THE ECONOMICS OF PATENTS

Property rights are a social construct that confers exclusive rights to a specific individual to use a specific asset. The intellectual property rights regime, in particular, provides inventors and their assignees with exclusive rights to their inventions as an incentive to private investment to undertake R&D and to commercialize new technologies or products. The agri-food industry has recently been a testing ground for the extension and management of new property rights. In particular, this extension of the legal rights of inventors has thereby extended the modern, market concept of invention to many countries where collective and traditional practices either undervalue or spurn individual enterprise and ingenuity, generating a significant debate about access and benefits sharing.

The root of the problem is the nature of invention and innovation. A variety of economists through the years (e.g. Plant, 1934; Schumpeter, 1954; Wright, 1983) have noted that perfect markets with free flow of information can be inimical to private investment in research. When the inventive step behind new products or processes can be codified into disembodied recipes or instructions (often called intellectual property), the knowledge often can be relatively easily disseminated to others skilled in the art. It is often impossible for inventors by themselves to exclude others from using the new ideas. This is the root of the economic problem. If a firm were to invest to create an invention under these conditions, any resulting benefits or profits would be bid away by imitators who would adopt or emulate the new invention, without compensating the firm or inventor for the sunk, fixed investments they made in the process of invention. This would make it impossible for innovators to recoup their investments. In the absence of any ability to stop or limit people from using a new idea, or ability to make them pay for the investment, for-profit investors would be unlikely to invest. A perfectly competitive market economy would then suffer a public good market failure due to inadequate investment in innovation.

This has been a particular problem in the area of germplasm and plant variety development, as incremental improvements are often self replicating, so that the benefits of inventions are easily distributed and appropriated by non-inventors. Farmers, government and industry have used a wide range of strategies over the years to overcome this problem. Farmers both in indigenous communities and in specific product lines have at times pooled their resources and collectively invested in sustaining or improving their germplasm or to develop new plant varieties. Similarly, government at times has used general government revenue to invest in developing new plant varieties, supporting research programs run by farmers, universities or at times private breeders. Industry also has invested selectively in new plant varieties. At times firms created closed environments (e.g. plantations) where they improved crops but used a mixture of biological and physical barriers, contracts and economic might to exclude others from using their improved varieties. Other times firms have invested in crops or species that had built-in protections, such as hybrid crops (e.g. corn) or crops that could only be reproduced asexually through cuttings or grafts. In spite of these efforts, there is substantial evidence that too little investment was being made in plant variety improvement (Alston et al., 1995).

More recently, governments have extended private property rights to the intellectual property embodied in new technologies and new plant varieties in an effort to close the investment gap in the agri-food world. The US and Europe have been the leaders in extending property rights in this area. As early as 1935 the US Patent Act provided plant patents for new asexually produced plant varieties (e.g. fruit tree varieties produced by cuttings and grafting). During the 1930s and 1940s a number of European countries proceeded to develop national plant breeders' rights rules for sexually-reproducing (open or self-pollinating) plants, and in 1970 the US followed suit with the Plant Varieties Protection Act. The system began to be internationalized with the 1978 negotiation of the International Union for the Protection of New Varieties of Plants (usually referred to as UPOV, an abbreviation based on the initials of its name in French: *Union pour la Protection des Obtentions Végétales*). While breeders were thereby given limited duration (mostly 18 years) rights to their varieties, these new rights were limited by two exemptions: farmers have the right to save seed from protected crops and use them to plant future crops, without need for a license or to pay further royalties or fees; while researchers have the right to use the protected germplasm for R&D purposes.

Beginning in the 1970s, there was a rapid encroachment of patents on the breeding system. During the early 1970s, a number of technologies related to gene spicing and transgenic manipulation were developed and protected under the provisions for general utility patents. When the first transgenic construct—an oil-eating bacteria—was presented in a patent application in the US, it was initially rejected. In 1980 the US Supreme Court ruled in *Diamond* v. *Chakrabarty* that the US patent law provides for patenting life-forms. In 1985 the first US patent for a sexually-reproducing living plant was issued. Since then, a wide range of technologies and plants have been patented. These patents provide additional protection over plant breeders' rights in that utility patents do not provide for any research exemption or for any farmers rights.

While this industrial concept of invention in the agri-food sector, and the corresponding rights and obligations, were initially restricted to the United States (as patents are only national instruments), it has been internationalized through a number of institutions. Many individual countries (such as Canada and the EU) have simply adapted and adopted many of the US policies to their system. But there has also been a multilateral effort to extend property rights internationally, through a range of international conventions related to patents, trade secrets, copyright, trade marks and various other property mechanisms, all coordinated through the World Intellectual Property Organization. While those voluntary processes had some effect, the pressure to conform increased in 1995 with the negotiation of the World Trade Organization Agreement and its subsidiary TRIPs Agreement. That bundle of agreements requires all member states (149 countries in December 2005, with another 29 countries negotiating to join) to grant patent protection to the full array of human inventions or, in cases such as plants, other *sui generis* or purpose built system of protection for plant inventions. Given that all member countries are now bound by the TRIPs rules, including developing countries which were given until 2006 to conform, the pressure is now on virtually every country to provide a full array of protections for invention.

While patents are often the preferred means of protecting intellectual property, in practice there are four classes of mechanisms used by innovators to protect their inventions. First, in addition to patents, there are a variety of other legal mechanisms, including trade secrets, plant breeders' rights, trademarks and copyright. Second, regulatory regimes established to protect public health and safety also effectively protect private property, by assigning rights and obligations to the owner and thereby impeding imitators. Similarly, a variety of commercial strategies can be adopted by companies to backstop their rights, including linking their inventions to protected complementary technologies, engineering in environmental constraints or using private contracts. Finally, some inventors have chosen simply to publish their findings, protecting their claims and access while precluding others from patenting or otherwise exclusively appropriating new inventions.

These mechanisms, by their very nature, do two things. First, they are designed to influence the rate of investment, invention and generation of economic value. Second, by their operation, they have the potential to change or restrict access to protected technologies and germplasm and can affect the distribution of benefits and costs. The literature offers a number of observations that are useful for informing the larger debate about this apparent trade-off.

The creation of private intellectual property rights for agri-food innovations in the past 20 years would appear to have spurred (or at least coincided with) increased private involvement. In the first instance, there has been extensive use of patents as a means of protecting research results—both the public and private sectors have moved aggressively to protect their innovations and intellectual property through patents. Similarly, since the advent of private intellectual property rights in the agricultural biotechnology industry in 1980, there has been a massive acceleration of private investment into agricultural applications, but only limited growth in public investment (Fuglie et al., 1996; Phillips and Khachatourians, 2001). The estimate is that the private sector contributed about 70% of the $4.4 billion of R&D expenditures undertaken in 2001, and that about 96% of all investment (public and private) was focused on developed world markets (Traxler, 2003). The vast majority of that investment was targeted on a narrow range of industrial crops now subject to patents. In contrast, even though developing countries made up 39% of the global seed market in 2001, only about 4% of the R&D expenditures on crop biotechnology were targeted there—coincidentally, IPRs are viewed as relatively weak or non-existent in those markets.

Ultimately, the question is whether the patents are creating new value, and who is gaining that value. The recent extension of legal IPRs—plant breeders' rights after about 1970 in North America and patents after 1980 in the US and elsewhere—is correlated both with some gains in yield and with the introduction of a range of new varieties with different agronomic or end use traits. A number of recent studies of the economic impact of biotechnology (e.g. Kalaitzandonakes, 2003) suggest that gross returns to these investments have been significant but, while estimates show that inventors and firms in the biotechnology industry are capturing between 12% and 57% of the total returns, there is limited evidence yet that the economic return to the investments is excessive (or even adequate). Thus, while the evidence does appear to show that patents generate more activity, the absence of any counterfactuals limits our ability to say whether we have absolutely more inventions and greater social welfare than we might otherwise have had.

Although most economists concur that intellectual property rights are an appropriate institutional response to overcome the public good market failure that would otherwise exist, there is significant debate about whether the system is operating to achieve optimal benefits.

Private property rights in theory should enable firms to exploit and benefit from their innovation, but in practice the fragmentation of the rights poses serious threats to both private and public benefits. One of the most pressing issues for many companies is the "freedom to operate" in a world of overlapping and interwoven claims to intellectual property. Both *de jure* (patents, Plant Breeders' Rights, trademark and trade secrets) and *de facto* (e.g. protected through contracts or via technical barriers such as hybrids) property rights create potential difficulties for adoption and diffusion of innovations. The absence of a clear "research exemption" for patented technologies (especially in light of *Madley* v. *University of North Carolina*, which concluded that UNC did not have the right to undertake university-based research using a patented technology without license) and the uneven patenting of various breeding tools and materials has raised particular concerns about the ability to develop and commercialize new plant varieties.

Finally, there is rising concern that the operation of the patent system is generating significant inequities in the global economy. The numbers tell a convincing story. The investment in research and the resulting technologies and new varieties are disproportionately focused on developed-world markets. Furthermore, an estimated 95% of the patents and plant breeders' rights in force today are issued in and assigned to inventors in developed countries (Sachs, 2000). In the end, the bulk of the new plant varieties commercialized are first, and often only, available in advanced industrial economies. Overall, economists and lawyers, while acknowledging the bias of the current system, tend to justify the current property rights systems as a second-best approach to accelerating investment in research, development and commercialization. The argument is that while the resulting monopolies may generate short-term profits, they ultimately raise the level of productivity and social welfare of a wide range of people over the long term—the generally accepted conclusion is that long-term gain justifies the short-term pain. But this skewed distribution of outputs and benefits worries many governments and policy advisors. They can often convincingly point to the important contributing role of genetic resources or traditional knowledge from developing countries and express serious concerns that unless some resolution is found, these resources may not (and maybe should not) be available as inputs to the global agri-food research system.

Three points should be kept in mind as you read the rest of this volume. First there are many different types of IP regimes. The more they overlap and interlock, the more likely that changing one mechanism will have little or no effect on the access or distribution of benefits of a new technology or product. Second, keep in mind that intellectual property rights are a two-edged sword—they offer short-term, limited monopoly profits for innovation in exchange for open access to proprietary ideas over the long-term. In the long-run, patents expire and proprietary knowledge enters the public domain, with a concomitant easing of access and wider distribution of benefits to consumers. Third, new inventions related to the life-science world are generating new wealth, but so far there is limited evidence that there are any large windfall gains being made. In short, the economic case for either maintaining the current system, or alternatively for changing the system, has not been convincingly made.

THE STRUCTURAL POWER OF KNOWLEDGE

Knowledge is a vital ingredient for success in information technology, biotechnology, genomics, nanotechnology, computer engineering, medicine, pharmaceuticals and other technology-driven industry. Prior to this period, a large share of knowledge in these sectors (especially for agricultural products) was in the public domain, often referred to as the commons. This is no longer true, as even products that are spin-offs from publicly funded research are often patented and exploited commercially. Most countries have changed their policies and relevant laws to allow universities to patent and licence inventions from government funded research (Eisenberg, 1987; Hoffmaster, 1993).

Private enterprise is currently leading the scientific breakthroughs in the knowledge industry due to the large volume of funds it is investing in R&D. The era when the state wholly-funded R&D through universities and public institutions appears to be over. Often, there is collaboration by multiple players—corporations, government and universities—for their own mutual benefit. This type of cooperation has been typified as a "triple helix" relationship (Etzkowitz and Leydesdorff, 1995). In order for this system to work, the state has to provide a favourable regulatory environment on which either the triple helix arrangement, or the private sector corporations alone, can carry on with R&D, in order to reap the benefits of such research.

A nurturing environment is required at both the national and international sphere. Corporations exert pressure on states to obtain a trade-friendly institutional support. This dependency of corporations on states is an indication that the primary function of regulation and ability to influence international affairs rests with the state (Gilpin and Gilpin, 1987). Nevertheless, it does not diminish the relevance of corporations in the emerging world order, particularly with respect to trade liberalization and appropriation of knowledge for commercial gain. The relationship between states and corporations in this regard can be likened to what Strange (1994, 165) described as a "complex and interlocking network of bargains that are partly economic and partly political."

Porter (1990) and others have asserted that knowledge influences competition in global markets. Strange's (1994) classification of knowledge as one of the four "structural powers" in International Political Economy (IPE) provides a basis for analyzing the full import of knowledge and the interests at play within the WTO/TRIPs and CBD paradigms. According to Strange (1994, 30): "Knowledge is power and whoever is able to develop or acquire and to deny the access of others to a kind of knowledge respected and sought by others; and whoever can control the channels by which it is communicated to those given access to it, will exercise a very special kind of structural power."

Prior to 1995, knowledge associated with improvements in Plant Genetic Resources (PGRs) was freely accessible to "the taker" (Low, 2001, 323). Low explains that this idea was inaugurated during the age of exploration, when researchers and travellers,

> [T]ransported plant species back to their own countries as new foods and raw materials for plant breeding. During this period, there was no formal proprietary right over knowledge or invention associated with PGRs. Thus, knowledge in traditional medicine, biodiversity and plant varieties resided in the community and not in an individual. There was no cost to its appropriation as the community had access to it, and freely transferred the knowledge from one generation to the other. However,

this notion changed with the coming into force of the TRIPs Agreement in 1995, coupled with the profits made by multinational pharmaceutical corporations from utilizing PGRs or the traditional knowledge about their uses.

It is widely acknowledged that Multinational Corporations (MNCs) in OECD countries played an important role in the build-up to, and throughout the negotiation process at, the Uruguay Round, to elevate Intellectual Property Right (IPR) onto the trade agenda of the General Agreement on Tariffs and Trade (GATT) and now the World Trade Organization (WTO). The result of this unprecedented effort was the TRIPs Agreement, which WTO administers (Drahos, 1999). Sylvia Ostry (1997) argued that the structure and function of the TRIPs Agreement is a direct result of a group of industry lawyers from the US engaging directly in the drafting of the agreement. Additionally, according to Drahos (1999, 429):

> Individual countries faced complete encirclement on the intellectual property issue. At the bilateral level it became a condition of any trade agreement with US and in the multilateral trade talks individual countries were faced by a US built consensus amongst the major trading powers on the issue.... And while talks at various levels were occurring, countries faced the menace of the US 301 process. The US and US business succeeded in their intellectual property objective because they pushed the issues relentlessly at all possible levels, in all possible fora, using all possible agents.

Vandana Shiva (1999, 158) captured the sense of outrage about the role of these corporations in the making of the TRIPs Agreement when she said: "the TRIPs Agreement of GATT is not a product of negotiations. It has been imposed by transnational corporations on the citizens of the world, by manipulating the governments of industrialized countries. The framework for the TRIPs Agreement was conceived and shaped by the Intellectual Property Committee (USA), Keidanren (Japan) and UNICE (Europe)."

Undoubtedly, the TRIPs Agreement revolutionalized worldwide ownership of intangibles, especially in non-OECD countries. While most developed countries had quite sophisticated IP systems, many developing countries had little or no policy and few processes to handle domestic or imported IP. Due to the binding nature of WTO, together with its "winner-take-all" structure, every country that signed onto the WTO now assumes the benefits and burdens arising from the TRIPs Agreement. As such, WTO member states have been required, and in compliance took steps, to redefine their national IPR regimes to conform to the standard of intellectual property regime prescribed in the TRIPs Agreement (van Wijk et al., 1993). The minimum IPR standard favours a much more robust institutional protection for patent holders than existed under the hitherto international conventions on intellectual property.

Prior to the WTO, IPRs were not within the domain of multilateral trade negotiations. Instead, the rights over intangibles were governed domestically, at times supported by international conventions such as the Berne Convention for the Protection of Literary and Artistic Works, 1971 and the Rome Convention (otherwise known as the International Convention for the Protection of Performers, Producers of Phonograms and Broadcasting Organizations, 1961). The challenge was that these conventions lacked an overseeing multilateral institutional body like the WTO. The conventions merely emphasized the principle of "national treatment" and relied extensively on the best efforts of ratifying states for compliance. In contrast, TRIPs enjoys the enforcement mechanism of the WTO; as part of the umbrella agreement, any derogation from the commitments in TRIPs can be penalized through sanctions on other aspects of trade, making the penalties much stiffer. In addition, the United States monitors compliance, applying trade sanctions against countries that fail to entrench and enforce the minimum standard of the IPR regime of the TRIPs Agreement. Acknowledging the huge influence US pressure had at the early stages of the negotiation and operation of the TRIPs Agreement, van Wijk et al. (1993, 19) contend:

> Apart from the multilateral route, the USA, and to a lesser extent the EC, have put bilateral political pressure on individual countries to strengthen the legal protection of advanced technologies, including biotechnology...The threat of sanctions and suspension of technological cooperation have had much greater impact than the IPR negotiations in WIPO or GATT.

KNOWLEDGE STRUCTURE UNDER THE TRIPS AND CBD

Knowledge is commonly recognized as an intrinsic yet abstract good that has the potential to enrich whoever is able to acquire and control it (Strange, 1994). Ultimately, it provides a competitive advantage to its holders. North (1991, 109) argues a "trader would invest in acquiring knowledge and skills to increase his wealth." Furthermore, the control of knowledge by the private sector often results in the alteration of the "basic institutional framework." The emerging institution, often a product of legislation, must be capable of adequately protecting the rights and interests of these private entities. In the modern society, the protection is often in the form of intellectual property rights granted for a certain period of time.

Drahos (1999, 429) argues that the essence of patent protection is to "give the owner the power to determine the physical reproduction of that object" over which patent exists. It guarantees the patent owner the enjoyment of proprietary benefits accruing from invention. It is also argued that patents encourage disclosure which otherwise will not be possible (Sell, 2000). Beside the issue of monopoly enjoyed by a patent holder during the duration of his patent, there are benefits for persons granted, albeit for a specific period of time, private ownership of knowledge. Under this arrangement, knowledge loses its public good attribute and becomes a private commodity, for persons who can afford it.

This is the issue with PGRs : multinational seed and pharmaceutical corporations have been able to obtain patents at the expense of source owners, tenderers and indigenous users of PGRs. One could argue that the patent system under TRIPs is skewed in favour of the technology-rich countries in the West. For instance, at the time of the Uruguay Round of multilateral trade negotiations, only the OECD countries had effective patent systems. Indeed, the United States before 1985 proactively amended its patent law to accommodate the demands of its Intellectual Property Committee (IPC). It was this new initiative that United States lobbied for and incorporated into the TRIPs Agreement.

What then is invention? From contemporary literatures on IPRs, invention must be capable of patent protection. In this regard, invention comprises three qualities—namely novelty, inventive step, and the capability of industrial application. Only an invention that meets the above criteria qualifies for patent protection under TRIPs. Member countries of WTO are required to protect such rights in their territories to enable holders of patents on any invention to attempt to recoup their R&D costs and to earn a profit on their efforts.

The above position differs fundamentally from the concept of knowledge within indigenous societies, especially those of developing countries. In these communities, knowledge is perceived as a public good that is accessible for personal and communal use. They regard any research—"scientific", traditional, or otherwise—to be publicly funded for the benefit of citizens. "Knowledge" for them must remain within the public domain, and incapable of private ownership or appropriation with a view to commercialization or monopolization of any sort. A person who, for instance, discovers any new method of improving a hitherto existing traditional knowledge is allowed to enjoy such discovery but without excluding other community members. The concept of excludability, which is one of the hallmarks of the liberal idea of property ownership, is absent in the traditional communal property paradigm.

Traditional knowledge or folklore lacks the basic Western criteria of invention for patent protection as provided in Article 27.1 of the TRIPs Agreement. Although a traditional process or product is based on invention, it is nevertheless (erroneously) classified as "ancient heritage" without an element of "novelty". Shiva (1999, 164) argues that "since indigenous medical systems are non industrial but part of the folk traditions or small-scale production processing and use, they also do not meet criterion" of industrial application. In their analysis of features of the traditional vision of knowledge, Cohendet and Joly (2001, 65) argue: "the traditional vision of knowledge production was characterized by a simple dichotomy: on the one hand, the distinction between science and technology which is logically deduced from the vision of a linear model of innovation; and on the other hand, the distinction between private and public research which is related to the fact that science was considered a public good."

Ascribing the characteristics of public good to knowledge is evidence that it could be used without reducing its value. According to Mytelka (2002, 43), "this meant that knowledge could diffuse widely if it were made public, something that was not always the case where mercantilist thinking prevailed."

At the Uruguay Round of multilateral trade negotiations there was a serious debate between developed and developing countries on the extent the proposed IPR regime should go in protecting traditional knowledge. Due to the power imbalance in negotiations, the outcome of that round—the TRIPs Agreement—did not protect traditional knowledge and folklore. The Agreement disregarded the "informal contribution of indigenous peoples and farmers to the maintenance and development of genetic diversity through years of cultivation and husbandry" (Lane, 1995). The outcome was a huge win for MNCs in the United States who utilize PGRs from the South as raw materials for their products. Under the TRIPs Agreement, biotechnology companies could exploit and use these PGRs in their improved status without any obligation to compensate either the source states from which these resources were obtained or the traditional communities that hold the existing traditional knowledge about their uses. In the process, the CBD's provision on access and benefit sharing becomes inconsequential within an IPR regime that is recognized and endorsed by WTO member states. Low (2001, 323) put the issue in context, asserting that "the increasing importance of plant biotechnology as a determinant of international competitiveness has helped to foster a vigorous challenge by developing countries to the concept of free accessibility to the PGRs (that is Plant Genetic Resources)."

PRINCIPLES OF ACCESS TO GENETIC RESOURCES AND BENEFIT SHARING

As already noted, the CBD provides the basic framework on access to genetic resources and the benefit sharing principle. The Convention perceives this principle as essential for the conservation and sustainable use of biological resources and the associated traditional knowledge. Consequently, it recognises the sovereign right of states over biological resources within their territory. It also acknowledges that each state has the authority to determine how to implement the regime of access to its genetic resources and the traditional knowledge about their uses. Specifically, Article 8(j) of the CBD establishes the principle of "equitable sharing" of benefits arising from the utilization of traditional knowledge, innovations and practices of indigenous and local communities. Article 15 further entrusts the source state with the power to negotiate and oversee that a favourable benefit sharing policy is established to implement this scheme. Indeed, Article 15(7) empowers each source state to use necessary legislative, administrative or policy measures to achieve this objective. However, Article 16(5) warns that in carrying out such initiatives, parties "shall co-operate in this regard subject to international law in order to ensure that such rights are supportive of and do not run counter to its objectives."

Notwithstanding that the CBD predates the TRIPs Agreement, the subordination of the CBD to other international laws appears to position it under the umbrella of the TRIPs Agreement. This is because the TRIPs Agreement qualifies as an international instrument within the scope of Article 16(5). Its objectives are private sector driven, unlike the CBD that is public sector oriented. The CBD, through Article 16(5), does not deal with questions of IPRs, leaving such issues to existing or future international agreements. Most recently, the 2001 WTO Doha Ministerial Declaration has taken up that challenge (known as the Doha Development Agenda) in the multilateral trade negotiations under the aegis of the WTO that have been underway since then.

THE STRUCTURE OF THE BOOK

The policy debate can be distilled down to a number of specific issues about the construction, operation and impact of patents on traditional knowledge. The most interesting aspects of the debate relate to the ultimate goals and impacts of patents, especially as they relate to concerns about access and benefits sharing.

Multiple goals are ascribed to patents. Patents are man-made constructs (some would call them "institutions") designed to achieve certain goals. As such, it is perfectly reasonable to debate the goals that patents are intended to support. This is where the debate gets difficult and divisive. Some see patents as simply one way of protecting or continuing the dominance of traditional power elites. It is generally accepted that patents differentially benefit owners and users and, can at times, disenfranchise others from some of the benefits or access to new ideas. What is less clear is whether this supports elites. Some argue that patents are actually subversive, often undercutting power elites by enabling inventors with new products or technologies to enter and compete with existing monopolies and oligopolies. Nevertheless, the dominant view is that, regardless of whether patents have the effect of sustaining elites, they were originally intended to (and many would continue to argue) generate economic and commercial progress through invention of products, processes and technologies and the adaptation, adoption and use of these new ideas. Thus the overriding focus in the debate tends to be on efficiency and growth, and not directly on equity or distributional issues. If equity is addressed at all in the dominant literature, it relates to the impact of patents on successive invention and innovation (i.e. questions of access) and discussion about the distribution of any concomitant benefits or costs of new technologies. While investigations of these matters can at times challenge the structure and operation of the patent system, they have not yet raised incommensurable concerns about the role of private initiative in the innovation system.

Part two of the book includes five chapters that lay out various aspects of the debate over sharing the benefits of inventions, PGRs and traditional knowledge. Genetic research has been criticized as being simply another mechanism for the developed world to benefit at the expense of the developing world. Genetic knowledge-gathering is frequently derided as neo-colonialism, bio-colonialism or biopiracy. In the context of non-human biological materials, the CBD creates a regime that provides national control over genetic resources in that country. In essence, the process created by the CBD requires that prior informed consent must be granted to the researchers by the host country on mutually agreeable terms which may, but need not, include benefit sharing provisions. There is no similar instrument that considers benefit sharing with regard to human biological materials. In the context of human genetic material a growing corpus of international legal instruments and ethical statements suggest the possibility of the emergence of an ethical and legal obligation to share the benefits of human genetic research with the populations from whom the bio-materials are obtained. While some indigenous groups and countries have supported these efforts, others have criticized them as being based on inappropriate notions of proprietorship. These benefits of human genetic research include, among other things, monetary benefits.

Onwuekwe (Chapter 2) opens the investigation with questioning the erroneous notion widely held in most biotechnology advanced countries (and supported by some international instruments) that PGRs *in situ* and *ex situ* (in gene banks) and the associated traditional knowledge (TK) on their uses are the "common heritage of humankind". He concludes that there is no legal or moral reason for designating indigenous plant varieties together with the TK about their uses and utility as part of the commons. He argues that such appropriation works against the interests of farmers in developing countries and, somewhat counter-intuitively, actually makes access less certain for breeders and farmers in developing nations. Phillips (Chapter 3) looks at the explicit system of farmers' privilege to see how the balancing act in international law and national plant breeders' rights systems provides special status for farmers. Using the *Monsanto Canada Inc.* v. *Schmeiser* case, Phillips concludes that the recent extension of new private rights to inventions related to seeds and the resulting complementarities between privately developed germplasm and industrial chemicals have caused new integrated relationships to emerge among biotechnology seed developers, chemical companies and farmers, which has effectively limited or removed the traditional privilege farmers have had to save and replant commercial seed.

That appears to be the end of the convergence of views. Most of our authors who directly address the question of benefits sharing express serious concerns about the scope of the current debate and structures being used to further the more equitable distribution of benefits. Gold and Castle (Chapter 4) argue that there is no obvious argument of natural right or for compensatory justice. They suggest that if benefit sharing is desired, there may be a distributive justice basis for action, but that any such argument is no more and may be less compelling related to PGR than for other historical and current inequities. They conclude that there may be a useful transactional approach to benefit sharing.

Craig and Mgbeoji both examine the case for and against bioassays involving indigenous people. Craig (Chapter 5) examines the cultural, legal and institutional approaches to defining "traditional knowledge" and its attachment to identifiable communities. She notes that patents have raised a debate about the dividing line between "biopiracy", which would appear to be mere (re)discovery, and bioprospecting, which involves an "inventive" step. There would appear to be some movement internationally on this (e.g. the CBD and the International Treaty on PGRs) but progress is likely to be slow, as it may need to be case-by-case. Mgbeoji (Chapter 6) takes a much stronger line against the current structure of the international law protection of indigenous peoples' knowledge, arguing that it for the most part simply validates the omnipresent reality of biopiracy.

Part three of the book addresses the practical question of how we might go about implementing Access and Benefits Sharing (ABS) provisions in our current system. Khoury (Chapter 7) rightly points out that new technologies have the potential to destroy value for some (Schumpeter, 1954, called this the process of "creative destruction"). Khoury examined the *Hoffman and Beaudoin* v. *Monsanto Canada Inc. and Bayer Cropscience Canada Holding Inc.* putative class action suit to determine whether and how co-existence of GM and organic crops could be managed. This case, which fundamentally revolves around the question of whether innovators are responsible for resulting negative economic externalities, could be fundamental to determining how large the attributable net benefits will be for new products that destabilize markets. Sheremeta and Knoppers (Chapter 8) and Jorgé Cabrera Medaglia (Chapter 9) both examine practical ABS arrangements. Neither identifies large profits that can be redistributed. Sheremata and Knoppers offer illustrations of how access and benefits sharing might actually play out under our current legal structures and with our current array of technologies, specifically how the concept and practice of benefit-sharing has emerged in relation to human genetic materials. Medaglia, meanwhile, examines the array of partnerships related to bioprospecting in Costa Rica, offering an insight into the practical challenges of operating related systems.

Part four of the book includes a discussion by Onwuekwe and Phillips (Chapter 10) that offers some concluding comments on what this all means in the context of processes underway for the development of more effective and equitable access and benefit sharing in the new millennium.

REFERENCES

Alston J, Norton G, Pardey P (1995) Science under scarcity: principles and practice of agricultural research evaluation and priority setting. Cornell University Press, Ithaca, NY

Cohendet P, Joly P-B (2001) The protection of technological knowledge: new issues in a learning economy. In: Daniel Achibugi, Bengt-Ake Lundvall (eds) The globalizing of the learning economy. Oxford University Press, Oxford

Drahos P (1999) Global property rights in information: the story of TRIPS at the GATT. In: Drahos (ed) Intellectual property. Ashgate, Aldershot

Eisenberg R (1987) Proprietary rights and the norms of science in biotechnology research. Yale Law J 97:177–223

Etzkowitz H, Leydesdorff L (1995) The triple helix—University-Industry-Government relations: a laboratory for knowledge based economic development. EASST Review 14:14–19

Fuglie K, Ballenger N, Day K, Klotz C, Ollinger M, Reilly J, Vasavada U, Yee J (1996) Agricultural research and development: public and private investments under alternative markets and institutions. AER-735. USDA, Economic Research Service, May

Gilpin R, Gilpin J (1987) The political economy of international relations. Princeton University Press, Princeton

Hoffmaster B (1993) Between the sacred and the profane: bodies, property, and the patents in the *Moore* case. I Prop J 7:115

Kalaitzandonakes N (ed) (2003) The economic and environmental impacts of agbiotech: a global perspective. Kluwer, New York

Lane M (1995) Invention or contrivance? Biotechnology, intellectual property rights and regulation. Presented paper at the 2nd meeting of the conference of the parties to convention on biological diversity, Jakarta, Indonesia, November, available online: http://www.acephale.org/bio-safety/IoC-index.htm

Low A (2001) The third revolution: plant genetic resources in developing countries and China: global village or global pillage? International Trade and Business Law Annual, April, VI, 323–360

Mytelka L (2002) Knowledge and structural power in the international political economy. In: Lawton TC, Rosenau JN, Verdun AC (eds) Strange power: shaping the parameters of international relations and international political economy. Ashgate, Aldershot

North D (1991) Institutions. J Econ Perspect 5(1):97

Ostry S (1997) The post-cold war trading system: who's on first?A twentieth century fund book. The University of Chicago Press, Chicago

Phillips P, Khachatourians G (2001) The biotechnology revolution in global agriculture: invention, innovation and investment in the canola sector. CABI Publishing, Wallingford, Oxon, UK

Plant A (1934) The economic theory concerning patents and inventions. Economica 1:30–51

Porter M (1990) The competitive advantage of Nations. The Free Press, New York

Sachs J (2000) A new map of the world. The Economist, 355(8176): June 24, 81–83

Schumpeter J (1954) Capitalism, socialism, and democracy. George Allen and Unwin, London

Sell S (2000) Big business and the new trade agreements: the future of WTO. In: Stubbs R, Underhill GRD (eds) Political economy and the changing global order, 2nd edn. Oxford University Press, Oxford

Shiva V (1999) Appropriation of indigenous knowledge and culture. In: Drahos P (ed) Intellectual property. Ashgate, Aldershot

Strange S (1994) States and markets. 2nd edn. Pinter, London & New York

Traxler G (2003) IPRs in agriculture: farm level issues. Presentation to the pre-meeting workshop on IPRs in agriculture: Implications for seed producers and users at the American Society of Agronomy-Crop Science Society of America-Soil Science Society of America Annual Meetings, Denver, November 2. Accessed at: http://www.farmfoundation.org/projects/03-25Speakerpapers.htm

van Wijk J, Cohen J, Komen J (1993) Intellectual property rights for agricultural biotechnology: options and implications for developing Countries. ISNAR Research Report No.3

Wright B (1983) The economics of invention incentives: patents, prizes and research contracts. Am Econ Rev 73(4):691–707

PART TWO

SHARING THE BENEFITS OF INVENTIONS, PGRS AND TRADITIONAL KNOWLEDGE

CHAPTER 2

IDEOLOGY OF THE COMMONS AND PROPERTY RIGHTS: WHO OWNS PLANT GENETIC RESOURCES AND THE ASSOCIATED TRADITIONAL KNOWLEDGE?

INTRODUCTION

The notion that Plant Genetic Resources (PGRs) found within the frontiers of sovereign states and the associated traditional knowledge on these resources are commons resources compound the problem of determining the appropriate compensation mechanism for local and indigenous peoples who have ensured the conservation and communal improvement of these resources (Onwuekwe, 2004). Despite dissatisfaction with the "biocolonization" undertones of modern biotechnology, the South[1] has been unable to persuade biotechnology-rich countries to extend proprietary interests to both "raw" and "locally" improved PGRs, and the associated traditional knowledge in the same manner that "elite" commercial cultivars of the North are regarded as proprietary. Treatment of PGRs and the related traditional knowledge as part of the commons has nothing to do with the immutability of the property system but resembles a scheme designed to preserve the North's[2] control of trade institutions with the attendant free access of state-occurring PGRs.

This chapter adopts a two-pronged approach in its analyses. First, it questions the validity of the notional extension of the commons concept to plant germplasm found within the territory of a sovereign state. If the concept of sovereignty over resources still means control of resources within a nation's territory then why treat PGRs differently? The erroneous view, which is widely held in most industrialized countries of the North, that these resources found within the borders of states are part of the commons, implies that they are non-excludable. In other words, the source states have no proprietary right to determine or regulate access to these resources notwithstanding the provisions of the 1992 UN Convention on Biological Diversity (CBD), which as this chapter demonstrates, re-confirms state control of territorially-occurring PGRs and the associated traditional knowledge. Of interest to this chapter is what peculiar circumstances justify this persistent notion of the commons when these resources are found within state boundaries not in the acclaimed global commons, such as the deep seabed? Furthermore, whose concept of property is paramount in this situation considering the likelihood of conflict of culture especially with the way ownership rights in state-occurring PGRs and/or the associated traditional knowledge are exercised/held?

Second, although PGRs are tangible items, the traditional knowledge with respect to their uses is not. Therefore, can traditional knowledge on the uses and values of germplasm rightly be categorized as part of the commons and therefore freely available to persons within and outside the source communities? Does the free sharing of this knowledge amongst source communities, and occasionally with outsiders, exclude traditional knowledge from the attributes of property capable of IPR's protection?

It is on the above basis that I discuss and analyse the various theoretical perspectives on the concept of the commons, the sovereignty of nations over resources, peoples and things within each state's territory, and the politicization of intellectual property rights through the instruments of international institutions. The chapter contends that the persistent notion that PGRs obtained from developing countries, and the traditional knowledge of their economic, medicinal and spiritual uses and values are common heritages of humankind is unjust, inequitable, and legally incomprehensible within the context of property rights, and the principle of sovereignty in international law. It further submits that although PGRs may be outside the purview of Intellectual Property Rights (IPRs) because they may not fit strictly as "innovations" or "inventions" within the Western paradigm of patents, there is no basis to deny source countries some other proprietary interests in these resources (see Article 27.1 the Agreement on Trade-Related Aspects of Intellectual Property Rights (the TRIPS Agreement)).

Suffice to say that this chapter recognizes the interdependence of nations and regions on PGRs for food and other uses. Hence, the chapter is not about which region contributes more germplasm to the other or is more dependent on food crops from other regions. Rather, the chapter disagrees with the one-sided notion prevalent in developed countries that PGRs from the South and the associated traditional knowledge on their uses are common heritages while their counter-parts—the so-called "elite" cultivars or "scientific" breeds—from the North are property capable of protection. Under this erroneous belief, both *in-situ* and *ex-situ* germplasm from the South are susceptible to free riding by those who contribute nothing to their conservation. This chapter argues that this development is unjustifiable and further ignores the often-touted diversity of the world. A diversity that is worthy of some legal protection.

THEORETICAL UNDERPINNING OF PRIVATE AND COMMONS PROPERTY

Kevin Gray argues that not all resources are propertized because a great proportion of the world's human and material resources are "still outside the threshold of property and therefore remain unregulated by any proprietary regime" (Gray, 1991). By way of illustration, Gray refers to the upper stratum of the airspace as part of the original common resources. He submits that common resources of this nature are incapable of private ownership or propertization rather that they are freely accessible by all humankind.[3] Because property comprises a "bundle of rights" and not a mere "thing," Gray contends that the important question in trespass or unlawful appropriation is "whether the defendants had taken anything that might be regarded as the plaintiff's "property." It is his opinion that private proprietary interest in a resource is possible only if the resource is "excludable," otherwise such resources will be "retained in the commons." He avers that commons resources are incapable of private ownership because of their physical, legal and moral attributes.[4] For Gray therefore, "excludability" connotes the ability of "a legal person to exercise regulatory control over the access of strangers to the various benefits inherent in the resource" (Gray, 1991).

Rejecting the "bundle of rights" or the exclusion theories as the only basis for the concept of property, Adam Mossoff argues that the "integrated theory of property" offers a holistic approach to the concept of property (Mossoff, 2003). Although Mossoff's theoretical approach does not dispute the importance of the right to exclude, he nevertheless argues that excludability is neither the only characteristic nor the most fundamental to the concept of property. For him, other essential factors of property such as acquisition, use, and disposal are vital for an adequate description of this concept. He asserts that "unlike the bundle theory, the integrated theory maintains that the elements of exclusive acquisition, use, and disposal represent a conceptual unity that together serve to give full meaning to the concept of property." Consequently, it is his view that possessory rights are as important as ownership rights, which incidentally is the final chain in the category of property rights. Relying extensively on the works of Grotius (1925), Pufendorf (Carr, 1994)[5] and Locke (Tully, 1980), he contends that the historical attempt to fathom the concept of property has focused on the "substantive possessory rights—the rights of acquisition, use and disposal—and the *right to exclude is left as only a corollary of these three core rights*" (Mossoff, 2003). For instance, Grotius was of the view that "men, who are the owners of property, should have the right to transfer ownership, either in whole or in part. For this right is present in the nature of ownership" (Grotius, 1925). Mossoff sums up his argument thus:

> It is possessory rights that form the basic building blocks of property, and the right to exclude enters the picture at the point that property comes to play a role as a political and legal right in a social context, i.e., at the point that property serves its normative function in organized society and politics.

Mossoff's contention that excludability is not the core concept of property finds support in Richard Epstein's (2000) response on "what principles decide *which* individuals have ownership rights (whatever they precisely entail) over *what* things." While acknowledging that the concept of property is nothing without state endorsement, Epstein argues that possessory right is the root of title in both tangible and intangible proprietary paradigms. In another article (Epstein, 1994), he discusses private and common property through a review of the conceptual basis of Locke's discussion of property and ownership. He questions Locke's insinuations that private property is preferable to common property because the latter is "undesirable and unstable" (Laslett, 1960; Locke, 2001). Epstein suggests the use of a "balancing approach" in the form of "appropriate *marginal* adjustments" to ameliorate the shortcomings in Locke's propositions on two constraints, namely guiding against waste and ensuring that what is left for others is good. He then recommends that "appropriate marginal adjustments" should always precede any removal of property from the commons.

Suffice it to say that despite the merit in Epstein's criticisms, Locke was, first and foremost, a liberal interested in bolstering the power of individuals over the power of the state or the collective.

Justin Hughes (1988), reviews the two theoretical pillars of property propounded respectively by Locke[6] and Hegel[7] as a basis for a general understanding of American property institutions and intellectual property in particular. He contends that these two theoretical perspectives, although initially meant to justify physical property, also provide a justification for intellectual property. Hughes uses the following three propositions to explain how ideas can be propertized under Locke's approach:

> [F]irst, that the production of ideas requires a person's labour; second, that these ideas are appropriated from a "common" which is not significantly devalued by the idea's removal; and third, that ideas can be made property without breaching non-waste condition.

Consequently, Hughes argues that intellectual property is the "propertization" of talent through a reward mechanism enjoyed for a limited duration. He states that it is this "self-defined expiration" that is "perhaps the greatest difference between the bundles of intellectual property rights and the bundles of rights over other types of property."

Pablo Eyzaguirre and Evan Dennis (2003) contend, "property rights facilitate coordination and collective action among people and groups by assigning responsibilities that carry with them the expectation of a stream of benefits." It is their view that a functional institution enhances a thriving property regime. This it achieves through proper distribution and application of property rights to increase the "social benefits of target groups." They refer to how property rights have been used in the past to promote two types of conservation, namely "*in-situ* conservation" and "living gene banks." Currently, efforts to integrate public and private incentives for biodiversity conservation have seen the uses of "property rights to legitimatize usufruct and regulated use of gathered plants" amongst other things. Thus, the application of property rights "has different impacts on the distribution of plant diversity."

Peter Drahos (1996) defines intellectual commons as consisting of "that part of the objective world of knowledge which is not subject" to property rights or other kind of barriers. He argues that this definition "emphasizes the idea that the intellectual commons is an independently existing resource which is open to use. Open to use does not mean, however, that abstract objects in the intellectual commons are necessarily accessible. Moreover, the fact that an abstract object is not in the intellectual commons and therefore not open to use does not mean that it is inaccessible." Citing some examples, Drahos contends that accessibility to intellectual commons of a community depends on the ability of the commoner to read and understand the local language or signs of the commons.[8] He also declares that "intellectual property rights can take an abstract object out of the intellectual commons, but this does not mean that it becomes inaccessible. Competent and capable persons can still gain access to the object provided they pay the relevant licence fee."

From a natural resource perspective, Christopher Gibbs and Daniel Bromley (1989) are of the view that common-property resources or rights are different from the concept of common heritage of humankind. They contend that the former constitutes a "special class of property rights which assure individuals access to resources over which they have collective claim." Unlike the "commons," therefore, the word "collective" or "group" as used by Gibbs and Bromley refers to an identifiable community rather than as reference to the whole world without exclusion. Collective property resource is a unique arrangement, which they argue is peculiar to traditional societies. They submit that it is created "when members of an interdependent group agree to limit their claims on a resource in the expectation that the other members of the group will do likewise. Rules of conduct in the use of a given resource are maintained to which all members of the interdependent group subscribe."

For Garrett Hardin (1993), the commons concept refers to resources that are susceptible to improper management, overuse, and destruction. He contends that the concept encourages "taking something out of the commons" without a corresponding desire or responsibility to replenish. This attitude, Hardin argues, will lead to tragedy as a resource that once flourished may become extinguished because each user "seeks to maximize his gain." Hardin acknowledges that the effort to exploit the commons resources is perceived as rational behaviour and therefore acceptable to the society. Thus, illustrating the consequences of such rational overuse with herdsmen, he issues the following warning:

> Therein is the tragedy. Each man is locked into a system that compels him to increase his herd without limit—in a world that is limited. Ruin is the destination toward which all men rush, each pursing his own best interest in a society that believes in the freedom of the commons. Freedom in a common brings ruin to all.

Hardin concedes that the "logic of the commons" is not a new phenomenon but nevertheless asserts that it has only been partially understood without attempt at generalization.

Dietz Thomas et al. (2003) acknowledge the importance but nevertheless an "oversimplification" of Hardin's tragedy of the commons thesis. They found his claim that "only two state-established institutional arrangements" could solve the problems of the depletion of the commons untenable. Most importantly, they assert he "missed the point" when he failed to recognize that "many social groups, including the herders on the commons that provided the metaphor for his analysis, have struggled successfully against threats of resource degradation by developing and maintaining self-governing institutions." They submit that developing effective governance regime that sustains the earth's "ability to support diverse life" is tough, and indeed resembles a "revolutionary race." In the real world, no one rule fits all. This is especially as diverse cultures and local governance regimes have successfully shepherded the commons through evolving rules. While agreeing that only few settings in the world are characterized by all of them, they nonetheless point out the following conditions as prerequisites for effective commons governance:

> (i) The resources and use of the resources by humans can be monitored, and the information can be verified and understood at relatively low costs... (ii) rates of exchange in resources, resource-user populations, technology, and economic and social institutions are moderate; (iii) communities maintain frequent face to face communication and dense social networks—sometimes called social capital—that increase the potential for trust, allow people to express and see emotional reactions of distrust, and lower the cost of monitoring behavior and inducing rule compliance; (iv) outsiders can be excluded at relatively low cost from using the resource (new entrants add to the harvesting pressure and typically lack understanding of the rules); and (v) users support rule enforcement.

While some societies are good at self-governance of resources, others are not that endowed. The greater challenge, they say, "involve systems that are intrinsically global (e.g. climate change) or are tightly linked to global pressures" because most times, they require governance from the international down to the local levels. Dietz et al., submit that the problem with global and national environmental policies is their disregard for "community-based governance and traditional tools." No single method provides all the answers.

In view of the foregoing, what is the true character of PGRs occurring within the threshold of state territory and the associated traditional knowledge? Are they collective community resources subject to some form of proprietary interest or are they indeed common heritages of humankind, which are incapable of private ownership?

COMMUNITY INTERESTS IN COMMUNAL PROPERTY

In property parlance, "commons" is used in reference to properties or things of importance, such as resources, outer space, or air, that are available for general use by a group, living and yet unborn (Onwuekwe, 2004). When used with a specific noun, the meaning is then derived within the context of use. Thus, "communal property" or "communal resources" simply refers to property or resources jointly owned by a community or family. Land, in most parts of Africa—particularly in agrarian communities—is an example of property that is usually held by communal ownership (Utuama, 1991). Land related resources, such as cropland or pasture, economic trees, and sacred or spiritual plants are often held jointly in traditional societies, for the mutual benefit of the communal (common) owners. Under this arrangement none of the co-owners may unilaterally exploit for personal gain the communal property without the consent and approval of the others (Onwuekwe, 2004). It is usual to share or parcel-out such communal property, often land, amongst the communal owners, for planting or cultivation. The communal owners may entrust a chief, a family head, or a group amongst them to manage the communal resources (Battiste and Henderson, 2000; Posey and Dutfield, 1996). Confirming this system of ownership with respect to Nigeria, Lord Haldane stated in *Amodu Tijani v. Secretary of Southern Nigeria*:[9]

> [I]n every case the chief or the headman of the community or village or head of the family has charge of the land and in loose mode of speech is sometimes called the owner. He is to some extent in the position of the trustee and as such holds the land for the use of the community or family. He has control of it and any member who wants a piece of it to cultivate or build upon goes to him for it.

With respect to traditional knowledge, Graham Dutfield warns that in this analysis, the history of how such knowledge became part of the public domain should not be overlooked. He argues:

> Indigenous peoples have for centuries endured abuses of their basic human rights, and they still tend to be politically, economically, and socially marginalized. It would therefore be naïve to suppose that it has ever been normal practice for their knowledge to be placed in the public domain and disseminated, with their prior informed consent *and* with respect for their customary laws and regulations concerning access, use and distribution of knowledge. It can plausibly be argued that unconsented placement of knowledge into the public domain does not in itself extinguish the legitimate entitlements of the holders and may in fact violate them (Dutfield, 2001).

Based on the foregoing, and the nature of PGRs, it is difficult to classify germplasm and the associated traditional knowledge as part of the commons, whether under international law or otherwise. PGRs do not exhibit the characteristics of resources classified as the common heritage of humankind. On the other hand, as I argued elsewhere (Onwuekwe, 2004), describing these resources as common properties under international law is to categorize them as resources beyond national jurisdiction. Such a description is without basis and therefore untenable. Germplasm are generally found within the borders of states. PGRs are also a result of many years of manual and intellectual contribution, in addition to the ongoing conservation and local plant breeding input of indigenous people and farmers in source communities. Such organized activities coupled with identifiable management over state-occurring

PGRs are non-existent under the international law concept of common property resources. With respect to such PGRs and the indigenous knowledge on their uses, the relationship is more of communal ownership and not common property or common heritage of humankind (RAFI, 1996).

Battiste and Henderson (2000) also describe the nature of this system of communal ownership and relationship with respect to indigenous knowledge. They used the concept of trusteeship to illustrate the responsibility of those entrusted with the management and overseeing of communal resources. According to them:

> Indigenous knowledge is ordinarily a communal right and is associated with a family, clan, tribe, or other kinship group. Only the group as a whole can consent to the sharing of indigenous knowledge and its consent must be given through specific decision-making procedures, which may differ depending on whether songs, stories, medicines, or some other aspect of heritage are involved.

On this basis, communal ownership of property is not an anathema (see Harris, 2001). The problem is the inability of the diverse world to understand and accept diversity. Therefore, state-occurring PGRs should not be equated with open resources or resources with open access under the ambiguous concept of the commons. This is because unlike state-occurring PGRs, open access resources are not "containable within national or regional boundaries."[10] However, as Shiva argued, the meaning and value of biodiversity of which PGRs is a part has changed with the "emergence of new biotechnologies" (Shiva, 1997). For Shiva, "biodiversity exists in specific countries and is used by specific communities. It is global only in its emerging role as raw material for global corporations" (Shiva, 1997).

The current absence of a centralized decision-maker over germplasm exploitation in source countries is due to the non-recognition of source nations' overriding proprietary interest in their PGRs. This is in contrast to "elite" commercial crops over which the patent holder enjoys proprietary interests. Thus, when the Food and Agriculture Organization (FAO) of the United Nations extended the concept of "common heritage" to "elite" cultivars through its resolution 8/83 of 1983, the North led by the United States and speaking with one voice, objected to this development (FAO, 1983). Not surprisingly, the North's objection was to protect the economic advantage enjoyed by their seed companies (Sedjo, 1988).

Depending on the context, the concept of "common heritage" is somehow similar to common ownership. However, on a closer examination, a common heritage resource still has a different character. For instance, it connotes free accessibility unless a limit is imposed by an agreement (Brownlie, 1998; Heller, 1998; Aoki, 1998; Kelly and Michelman, 1980). In addition, sovereignty is not claimable by individual states over such areas or resources (Brownlie, 1998). Rather, like the high seas, which have the "character of *res communis*,"[11] common heritage resources are "open to the use and enjoyment of all states on an equal basis" (Brownlie, 1998).

The common heritage concept is similar to the concept of common-pool resources. According to Dustin Becker and Elinor Ostrom (1995), a common-pool resource has two main attributes—"the difficulty of excluding beneficiaries and the subtractability of benefits consumed by one person from those available to others." With respect to excludability, they reiterate that if the cost of trying to exclude potential beneficiaries from using the resources or consuming the goods is prohibitive then it is not worth it. In other cases, the potential gain from exclusion or restriction of use may be less than the additional costs from "instituting a mechanism to control use." As such, they acknowledge the importance of backing possible restrictions with "property rights that are feasible to defend (in an economic and legal sense)."

Becker and Ostrom also explain that the extent of "subtractability of one person's use from that available to be used by others" differs from one good or event to another. They illustrate the idea of subtractability with the catching of a ton of fish from the river and the enjoyment of sunset. While the former subtracts what is available to others to fish, the latter does not diminish or subtract from others' enjoyment of the sunset. They submit that goods with exclusion and subtractability attributes are often public goods.

STATE-OCCURRING PLANT GENETIC RESOURCES AND TRADITIONAL KNOWLEDGE

The contention of developing countries that PGRs occurring within their state boundaries are not part of the commons finds support in the treatment of natural resources found within a country's borders under international law. Generally, countries are regarded as the sovereign owners of natural resources within their territories. Medicinal plants, such as the turmeric plants found in India, and agricultural food crops, such as quinoa (*Chenopodium quinoa*)—a high nutritious drought-resistant food crop of which numerous varieties are grown in the Andean countries of Bolivia and Peru, are natural resources, and therefore within the juridical control of these source countries (Dutfield, 1999). In the same way, traditional knowledge and practices of the Indians on the uses of the neem or turmeric tree, and of the Bolivians or Peruvians on quinoa, are within the category of knowledge for which intellectual property protection should be extended.

It is inequitable to apply different laws on seemingly very similar circumstances. PGRs found within the borders of a sovereign state and the traditional knowledge about their uses are outside the category of resources or things currently classified in international law as part of the commons, such as resources in the deep seabed.[12] Hence, it is improper to liken PGRs occurring within the territory of a state to the deep seabed, the air, the moon, the continent of Antarctica (Antarctic Treaty, 1959), or the "outer space"[13] which are typical global commons that are incapable of private or communal ownership. To do otherwise is to undermine the concept of sovereign control of natural resources found within a state's territory.

Jack Kloppenburg (1988) disagrees with the persistent notion that PGRs are the common heritage of humankind. This is because, as he and Daniel Kleinman argue on another occasion, under the commons concept, germplasm resembles other common goods, which do not give rise to compensatory obligations (Kloppenburg and Kleinman, 1988). Kloppenburg claims that the commons concept with respect to state-occurring germplasm is an institutionalization of the historical relationship by which the capitalist core has continuously exploited the global periphery (Kloppenburg, 1988). He says that unlike developing countries, capitalist nations have known for a long time that "plant genetic resources are a strategic resource of tremendous value." It was this lack of knowledge that prevented developing countries from paying much attention to the free accessibility and exploitation of these resources by the biotechnology-rich countries until recently. However, with this recent knowledge "has come an awareness of the asymmetric character of gene flows in the world system and of the implications of political and institutional control over plant genetic resources."

Kloppenburg and Daniel Kleinman (1988), also canvass for a national sovereignty over state-occurring PGRs in line with Article 8(j) of the CBD. They acknowledge the huge capitalist interests backing private proprietary rights over "elite" crops, despite the refusal of these same interests to recognize the property rights of sources of origin to germplasm. Thus, they disagree with the widespread notion in Western paradigm that PGRs are part of the common heritage of humankind on the basis that such opinion falls short of the international recognition of sovereign rights over resources found within a country's territory. They also argue for a change in this "capitalist conspiracy" against developing countries. Their basis is that "law is a social creation, not an immutable reflection of the natural order." This connects to Kloppenburg's contention that the notion, albeit unorthodox "classification" of plant genetic resources as part of the commons is not because of the "essential character of the germplasm itself" but with its "social history and political economy."

Indeed, Eyzaguirre and Dennis (2003) point out that the value of PGR goes beyond its physical attributes. Rather, "much of their value is determined when engaged in a relationship with the ecosystem, farmers, and scientists." This is what helps farmers and others who use PGR to "adapt to changing environmental, market and social conditions." This is particularly useful to local cultivars that depend on it for most of their livelihood. As such, "by creating ownership rights" over PGR, property rights will "make long term investment in sustainability worthwhile and diversity an effective technology." They reject the extension of the commons resource taxonomy to PGR on the basis that "plant genetic resources are intrinsically different from other common property resources." This is because "the value of PGR is not in any single variety, but in the bundle of varieties, *each of which contains different characteristics from other varieties.* This value is not diminished due to the manner in which local communities collectively hold, improve and manage PGR.

Kloppenburg and Kleinman (1987) in their seminal empirical investigation of the roots of this controversy on the notion of common heritage ascribed to PGRs noted that both North and South are dependent on each other's PGRs for food crops and survival. Essentially, no part of the world is independent on plant germplasm. They acknowledge that the North uses more of the South's germplasm than it contributes to them, and therefore hugely indebted to the "global plant genetic estate." Kloppenburg and Kleinman set out their findings along food crops and industrial crops as most effective ways to measure genetic interdependence of nations or regions. With respect to food crops, they concluded that "the six regions that contain nearly all of the world's less developed nations (Chino-Japanese, Indochinese, Hindustanean, West Central Asiatic, Africa, and Latin American) together have contributed the plant genetic material that has provided the base for fully 95.7% of the global food crop production." Consequently, they concede that the description of the North as a rich but "gene-poor" recipient of the genetic materials from the poor but "gene-rich" South is proper. They also found that genetic dependence is not peculiar to the North. For instance, the African region's dependency index is 87.7% mainly from Latin American crops (maize, cassava, and sweet potatoes). With respect to industrial crop genetic geography, they discovered that the dependence is not along North-South divide but demonstrates an "interdependence of regions within each hemisphere." On this note, they assert:

> Similarly interdependence can be found throughout the South in a variety of crops and regions. Maintaining intrahemispheric flows of plant germplasm is as much a central concern with regard to industrial crops as maintenance of intrahemispheric flows is the central issue with regard to food crops.

Kloppenburg and Kleinman acknowledge that because the North is used to private ownership of their improved plant germplasm, they are unlikely to declare private breed seeds as public goods. Nevertheless, they regard as ironic, and somewhat inequitable, the notion that PGRs from the South are part of the commons available for free accessibility. They conclude that since "the concept of common heritage is unworkable for one category of plant germplasm, we do not believe that it should be maintained with regard to other categories." Consequently, they request the North to back down on their position of "unrecompensed access to a natural resource that is objectively valuable, even if difficult to value." To do otherwise, will conflict with the rights of states over the natural resources within their sovereign borders.

Fowler et al. (2001) echo the above position of Kloppenburg and Kleinman that nation states are dependent on each other for food crops. That although early crop transfers were somewhat uneven—from South to North—this anomaly seems to have been corrected with increased reliance by developing countries "on their own production of non-indigenous crops (from other developing country regions) to meet food needs." What Fowler et al., fail to address in this piece was the persistent notion that plant germplasm from the South are common heritages of humankind and not the property of the source country or region.

The CBD does not define traditional knowledge. However, it defines biological resources as "genetic resources, organisms or part thereof, population, or any other biotic component of ecosystems with actual or potential use for humanity (The Convention on Biological Diversity, 1992). In addition, Article 8(j) of the Convention mandates contracting parties to "respect, preserve and maintain knowledge, innovations and practices of indigenous and local communities embodying traditional lifestyles relevant for the conservation and sustainable use of biological diversity..."

Similarly, neither the Rio Declaration nor Agenda 21 defines "traditional knowledge," notwithstanding they both used the term in various forms. For instance, Chapter 26 of Agenda 21 requires national governments to take certain actions in empowering indigenous peoples through "measures that include recognition of their values, *traditional knowledge* and resource management practices..." (see Blakeney, 2001).

Marie Battiste and James Henderson (2000), claim that there is no "short answer" in the definition of indigenous knowledge. They reject attempts to pigeonhole indigenous knowledge, practices and cultures into the "precision and certainty" of Eurocentricism. They criticize Eurocentric scholars who are eager to "impose a definition" on indigenous knowledge and "when it fails to comply with any universal standard by deductive logic, quibble over its meaning." This attitude, according to Battiste and Henderson, is typical with the assumption of superiority of the Eurocentric worldview by its thinkers who also wish to "impose it on others." This, they assert is why "[t]he Eurocentric quest for universal definitions has raised suspicion among indigenous peoples, who do not want to be assimilated into Eurocentric categories." Elaborating the problems of this one sided approach in understanding a diverse world, they submit:

> The Eurocentric strategy of universal definitions and absolute knowledge has made its scholarship unable to know and respect indigenous knowledge and heritage. To attempt to evaluate indigenous worldviews in absolute and universal terms is irrational. Using Eurocentric analysis, one cannot make rational choices among conflicting worldviews, especially those held by others. No worldview describes an ecology more accurately than others do. All worldviews describe some parts of the ecology completely, though in their own way. No worldview has the power to describe the entire universe.

Notwithstanding the non-homogeneity of indigenous or traditional societies, and the difficulty attendant with blanket definition, Battiste and Henderson still define indigenous knowledge (which they also call "traditional ecological knowledge") as "the expression of the vibrant relationships between the people, their ecosystems, and the other living beings and spirits that share their lands." It is their claim that "all aspects of this knowledge are interrelated and cannot be separated from the traditional territories of the people concerned." They reject the extension of the commons concept to traditional knowledge on the basis that indigenous people have always regarded this knowledge as proprietary and confidential.

Ikechi Mgbeoji makes an interesting argument when he suggested that the concept of "plant resources-related knowledge" more appropriately describes the knowledge and practices of indigenous and source communities on the uses of their PGRs (Mgbeoji, 2002). He suggests that there is an inherent limitation in the commonly used terms "traditional" or "indigenous" in reference to the knowledge of traditional societies about PGRs. Mgbeoji contends that his suggested term, "plant resources-related knowledge," is inclusive and further erases the mistaken belief often associated with "traditional knowledge" as being "antiquated and inferior to Western science" as well as being "culture bound and ethnic in nature." He defines the concept to include knowledge of "the plant resource, parts or derivatives thereof and the knowledge of their various uses regardless of the particular paradigm in which it may be practised."[14]

Notwithstanding the objections of developing countries to this erroneous extension of the notion of commons concept to the state-occurring PGRs and the traditional knowledge about their uses, Alex Low submits that the real problem is with "establishing control and utilization" over PGRs (Low, 2001). For him, the greatest problem faced by developing countries is "knowing what the resources are, and how they can be manipulated." He concludes that the free taking of germplasm without compensation will continue because the "issue is not who owns the gene, but who possess the technology to make the gene useful" (see also Goldstein, 1988).

Low criticizes the Western individualistic conception of property because it "ignores the collective labour of generations" of germplasm source communities (Low, 2001). He points out that even "under Locke's labour theory of property, the developed country reasoning cannot be sustained because it overlooks the fact that PGRs from developing countries are not simple products of nature. In fact, labour has been expended over time by indigenous people and generations in preserving and improving these PGRs." This ignorance, he argues, is why PGRs are still regarded as part of the commons thereby denying communities that sustain these resources through biodiversity activities any benefit from the propertization of elite germplasm developed in Western laboratories from local cultivars (PGRs). Besides being inequitable, he believes the approach is a "cloak for reality" and a demonstration of Western superiority that belittles other cultures and property ownership structures that are not as individualistically driven. As he notes,

> [i]t advances the notion that PGRs are of no value until the intellect of industrial countries has been applied to them and thus lacks respect for other cultural perceptions of property. Above all, it is less concerned with the purported interests of mankind than with the interests of maintaining Western hegemony in the control and utilization of PGRs.

As an alternative, Low suggests that if communally improved PGRs are not the type of property for which intellectual property rights protection may be extended then another form of proprietary protection should be devised to enable source communities to reap the benefit of centuries of biodiversity conservation and improvement. Finally, he proposes the identification and use of either the "proprietary theory, the equity theory as modified by the concept of unjust enrichment, or the incentive theory" as a basis for compensating PGR source states.

BIODIVERSITY AND THE COMPENSATORY THEORY

Generally, source communities are required (or expected) to conserve PGRs for the benefit of humankind (both those living and the yet unborn).[15] Without an economic incentive to undertake this responsibility, it is doubtful if the international community, particularly the North, have a moral justification for this demand or expectation. It does not matter that these communities have undertaken such responsibilities including improving PGRs for better quality yields in the past without compensation. Suffice it to say that up till date, there is no single international treaty or instrument that classifies, categorizes, or designates PGRs occurring within the territorial boundaries of a state or the associated traditional knowledge as common heritage of humankind. The extension of the commons concept to these resources and the related traditional knowledge about their uses is a scheme crafted by the industrialized biotechnology countries to confine to the public domain some valuable resources that are found in other states but pivotal for their own biotechnological development.

The argument that agricultural biotechnology holds the potential for ending the food problems of Third World countries is more rhetoric than reality (Phillips, 2001; Murphy, 2001; Zilberman et al., 1998). Moreover, since agricultural biotechnology promotes monoculture rather than biodiversity, it essentially contradicts the global interest in biodiversity (UNCED, 1992; CBD (1992), which came into force on 19 December 1993). Some commentators are of the opinion that biotechnologies and the attendant encouragement of monocultures may cause serious economic and social problems for Third World agrarian communities (Gertler, 1998; Mgbeoji, 2001). These include loss of earnings from local crops, increased costs for patented seeds, pollution of farms with bio-engineered seeds (Greenpeace Canada, 2002) similar to some of the issues and arguments in the *Monsanto v. Schmeiser* case ([2001] FCT 256 and Shiva 1993). As Gertler noted, "women and men who cannot relate to the new paradigm will be increasingly alienated, contributing to a loss of cultural diversity in farming and agriculture" (Gertler, 1998).

Despite the economic and social implications of biotechnology and associated monocultures, there is no legal support under the current international jurisprudence for the denial of source countries' proprietary interest in germplasm. The current denial ignores the past and ongoing intellectual input of local farmers in the improvement of seeds and plant life forms occurring within their communities, albeit through non-modern biotechnology (or "scientific") methods (Kloppenburg, 1988). This non-recognition of property rights of source communities over their plant germplasm and the associated traditional knowledge also has far-reaching economic and social consequences for these societies (Busch, et al,1991). Admittedly, developing countries allowed the carting away of their exotic plants and genetic resources for storage, first in botanical gardens, and subsequently, in gene banks (Pistorius and van Wijk, 1999). Source countries' helplessness on this development were made worse because the initial exploitation took place during the colonial era, while the subsequent actions are occurring in the context of the present period of economic imperialism.[16]

Due to the special protection of "scientifically" improved plant varieties, first through plant breeders rights (UPOV, 1978; Plant Patent Act, 1952) and subsequently, by intellectual property rights (such as the US Plant Variety Protection Act, 1970; TRIPs Agreement, 1995), source countries of PGRs are contesting the continued treatment of their plant germplasm as the common heritage of humankind. This persistent notion has made it impossible for them to have intellectual property protection extended to local but communally improved PGRs or the traditional knowledge about their uses. Furthermore, even the concept of Farmers' Rights that the International Union for the Protection of New Varieties of Plants (UPOV)[17] originally promoted was discontinued by its deletion in subsequent amendments (see UPOV Amendment, 1996).[18] Developing countries therefore question the common heritage taxonomy ascribed to both raw and locally improved PGRs. They contend that it is inequitable and represents an attempt to extend the proboscis of colonial powers to their resources once again, albeit through capitalistically inspired international institutions (Krasner, 2000). Thus, besides the non-compensation of local farmers for the use of their PGRs, designating these resources as part of the commons is capable of dislodging indigenous farming communities in a manner similar to the dramatic impacts of the break up of "plantation economies" when rubber seeds were smuggled out of Brazil in 1876 by officials of Kew Botanical Gardens (Pistorius and van Wijk, 1999).

According to Vandana Shiva, the nature of knowledge and innovation recognized under the Agreement on Trade Related Aspects of Intellectual Property Rights ("TRIPs Agreement") ignores other "kinds of knowledge, ideas, and innovations that take place in the "intellectual commons"—in villages among farmers, in forests among tribespeople and even in universities among scientists" (Shiva, 1997). This, she argues, is because the intellectual property rights regime incorporated in the TRIPs Agreement is "based on a highly restricted concept of innovation" largely dependent on the United State's patent regime. Shiva refers to this development as an imposition of a monoculture of knowledge on poor nations of the world by the powerful nations of the North led by the United States. She further describes this one-size-fits-all assumption of IPRs as the perpetuation of the history of "inequality and poverty."

On whether the extension of IPRs to communally improved PGRs or traditional knowledge with respect to their agricultural and health uses will solve the problem of bio-piracy, Shiva disagrees with any such arrangement if the resulting economic structure will undermine or "displace the indigenous lifestyles and economic systems."

In contrast, Anthony Stenson and Tim Gray (1999), contend that traditional societies or centres of origin are not morally entitled to any intellectual property rights in their communally improved PGRs and/or the associated traditional knowledge. It is their opinion that arguments to the contrary lack support under the Lockean labour theory of property or what is often referred to as the entitlement justification theory. Explaining the operation of the labour theory of property in practice, Stenson and Gray assert that communities are "incapable of performing entitlement-creating acts," which in the Lockean sense involves an identifiable labour. This is because,

> [F]or entities to perform labour, they need to be in possession of many mental tools: a conception of the future, an idea of their good and of how to achieve that good, the knowledge of how to perform the act that will achieve their aim, and so on. In short, to labour, an entity has to be capable of executing a rational plan aimed at some end. It is hard to see how a community, even if it were an independently existing entity, could achieve such a thing without a brain of its own.

Based on the foregoing, Stenson and Gray conclude that only individuals who engage in labour rather than their communities are capable of owning property. They distinguish corporations from communities. They contend that corporations have legal personality that provides them the capacity to hold property notwithstanding they are dependent on the directing minds of individuals who occupy position of directors in them. But even with such legal personality, they say, there is no doubt that "ownership, and with it control in the final analysis, resides with the shareholders," who are individuals. Surprisingly, Stenson and Gray ignore the fact that in modern corporations, shareholding is not exclusive to individuals. Corporations may and indeed do hold shares in other corporations. After all, they enjoy the bundle of legal personality by which they can own properties or make investments in third party entities. Such an accomplishment in Stenson and Gray's parlance amounts to full, partial or shared "ownership and control" in the other entity or interest. Or would Stenson and Gray suggest that the veil of such holding corporations be lifted to find out its make up? If the latter is their argument, and their analysis so suggests, then this is the same way those behind communities, albeit states, can be seen as constituting it and labouring within it to achieve the best value for themselves. It appears that Stenson and Gray missed this analogous relationship.

It is on their above mistaken conclusion that Stenson and Gray contend that it would be wrong to attribute author—figures to a community since it lacks the ability to bind future generations with its current opinion. It becomes even more difficult, they reason, when it involves IPR claims over traditional knowledge even though it is not a single act of creation but "the result of centuries of collective experience," skill, practices and knowledge. This, they argue, falls outside the paradigm of individual acts for which IPRs are available.

With respect to the concept of the common heritage of humankind and its erroneous application to genetic resource control and traditional knowledge contrary to the reports of the World Intellectual Property Organization (WIPO), Stenson and Gray submit that there is nothing wrong with the notion or classification (Stenson and Gray, 1999). In fact, it is their view that the common heritage concept is compatible with the principles of community IPR and national sovereignty, respectively. They regard the concept as the underpinning for a benefit sharing arrangement enshrined in the CBD; by which the naturally endowed germplasm states are required to allow free access to the biotechnology-rich countries in the North in return for technology transfer, and some other mutually agreed benefits from their utilization of raw PGRs. Stenson and Gray rely on Rawls theory of justice to support their position and submit that genetic resources found in developing countries should be used for the common benefit of humankind in order to improve agricultural yields or seed varieties and health care systems. This is on the grounds that communities have no moral or natural rights to their traditional varieties since they have not "laboured to create them." They also point out that based on the entitlement theory of intellectual property that there is no double standard in providing IPRs protection to improved PGRs without a corresponding proprietary protection to "raw" germplasm and the related traditional knowledge.

Graham Dutfield challenges the above postulations of Stenson and Gray on various grounds. First, he submits that traditional peoples and communities perceive the globalization of the patent regulations applicable in the United States and Europe as a form of neo-colonialism (Dutfield, 2001b). Second, he disagrees with the criteria for patents[19] as the basis for non-extension of patent protection to Traditional Knowledge (TK). Accordingly, Dutfield asserts:

> No matter how novelty and non-obviousness are defined in patent laws, researchers and companies may be tempted to misappropriate TK, especially in those jurisdictions where patent office staff are known to have insufficient time or resources to conduct thorough prior art research and examinations.[20]

Third, Dutfield also debunks the notion of individual inventor as part of the reasons for not extending patent protection to traditional knowledge adjudged to be collectively held. He argues that unlike when individual flashes of genius were important for the granting of patents, the law and doctrine on patents have begun "to accommodate the collective notion of inventorship from as early as 1880, first in Germany and then elsewhere."

Dutfield contends that traditional knowledge exists notwithstanding that it suffers from a definitional dilemma. He suggests a broad description of the term as opposed to a specific definition because of its "incredibly diverse" nature "not just between different peoples, groups, and communities, but within them too." Accordingly, he adopts Martha Johnson's definition of traditional ecological knowledge, as a starting point:

> A body of knowledge built by a group of people through generations living in close contact with nature. It includes a system of classification, a set of empirical observations about the local environment, and a system of self-management that governs resource use (Dutfield, 2001a, 2001b).

For Dutfield, the above definition is an indication that there is some element of scientific basis for Traditional Knowledge (TK) notwithstanding its distinguishable uniqueness. Consequently, he states:

Some TK is, at least to some degree, scientific even if the form of expression may seem unscientific to most of us. For example, an indigenous person and a scientist may both know that quinine bark can cure malaria. But they are likely to describe what they know in very different ways that may be mutually unintelligible (even when communicated in the same language) (Dutfield, 2001b).

Notwithstanding his criticism of the current regime of intellectual property rights, Dutfield expresses his concern at the inability to trace the origin or sources of traditional knowledge for purposes of compensation or attribution of creativity rights and or benefits. Although he acknowledges arguments in support of collective intellectual property of traditional communities because of their "collective nature of creative processes," he nevertheless questions the validity of such generalizations especially with respect to cases where "two or more peoples or communities share the knowledge," or even where the author may be unknown.

Elinor Ostrom (1999, 2004) provides an illuminating account of the advantages and problems of collective action properties. She explains that collective action refers to when "more than individual is required to contribute to an effort to achieve an outcome." Unfortunately, these individuals are often unable to exclude free riders from benefiting from their collective action. She questions the efficacy of externally imposed property rights regime on collective action properties. In contrast, she proposes leaving the individuals to manage their collective action properties through indigenous or local institutions they have "established over time." Ostrom's proposal may work for common resource properties identified to be within the immediate control of indigenous or local communities. After all, the labour of these groups in tending and conserving natural resources—water, forests, fisheries, and rangelands—are vital and should not be upturned with modern institutions staffed with bureaucrats. Unfortunately, her suggestions are inapplicable to PGRs and the associated traditional knowledge because they are not common heritage resources. Their unique nature presents both the international and municipal institutions of property with a challenge to fashion a property rights' regime that recognizes past and ongoing labour of local communities over PGRs and their associated traditional knowledge. It is a task that deserves thorough accommodation by the different parties—North and South—in this debate about the proprietary characters of these resources.

As pointed out elsewhere in this book, the concept of modern intellectual property right enables an inventor to enjoy monopoly over an invention for a certain period of time. Notable forms of IPR include patents, copyrights, and trademark. With patents, three criteria must exist before an invention will qualify as patentable subject matter. These are novelty, inventive step and commercial viability. The idea of an individual inventor enjoying patent monopoly over an invention is virtually non-existent in modern societies. Corporations appear to own more patents than individuals, either through licensing or by funding the research that gave rise to the invention. In other words, the world is no longer fixated that only individuals can own or apply for patents over an invention. Similarly, nothing should preclude communities or collective owners from seeking protection of their ideas, products from those ideas and improvements of the products within the available property regime, such as the applicable IPR. It only becomes difficult to achieve if such ideas or products are classified as part of the commons.

STATE-OCCURRING PGRS UNDER THE CBD AND THE GLOBAL SEED TREATY

The CBD and the current Global Seed Treaty[21] are two international agreements that recognize the proprietary nature of state-occurring PGRs (Onwuekwe, 2004). They both acknowledge a source country's sovereign ownership of germplasm found in its territory. The ownership also extends to *ex-situ* germplasm originally taken from a source community for storage in gene banks located overseas. Although these agreements focus primarily on biodiversity and food security respectively, they nevertheless confirm that PGRs found within the territorial boundaries of a sovereign state, and the knowledge on their uses, belong to that state. Under these agreements PGR source states are equivalent to trustees, albeit in a loose sense. Despite this, the industrialized capitalist nations have conveniently, and without any juridical support, extended the commons concept to traditional knowledge of local farmers and indigenous peoples on the uses of their PGRs. This unilateral development appears to have found strong notional backing in some international institutions (Evenson, 1999) and amongst some Eurocentric scholars (Battiste and Henderson, 2000). There is no doubt that Western biotechnology countries desire this status quo to remain, because:

> [T]he ideology of common heritage and the norm of free exchange of plant germplasm have greatly benefited the advanced capitalist nations, which not only have the greatest need for and capacity to collect exotic plant materials but also have a superior scientific capacity to use them (Kloppenburg, 1988).

The fact that a sovereign nation is unaware that uranium deposits or any other natural resources lie within its borders does not make future control of such resources non-existent. (UN General Assembly Resolution 3129 XXVIII (1973) and Charter of Economic Rights and Duties of States, 1974). It is also of no consequence that a foreign corporation discovered the resources. In modern commercial relationships, the best any corporation may acquire in this instance would be a licence to exploit the resources subject to payment of royalty or resource rent. Consequently, a country's ignorance or lack of technology to exploit its natural resources has never been a disadvantage for its control over these resources. State-occurring PGRs should not be different or else it will amount to a forceful appropriation without any support in law.

COMMODIFICATION OF TRADITIONAL KNOWLEDGE

Ascribing monetary value to traditional knowledge for purposes of compensating the source communities from whom such knowledge is derived may help in meeting the core tenets of Article 8(j) of the CBD. However, this suggestion is not absolute as other compensatory mechanisms may be agreed with each source community that could still meet the CBD requirements. The main focus is that the benefits from biotechnology gains for the uses of PGRs and the associated traditional knowledge should accrue also to the source communities. After all, it is them who labour to preserve and conserve the biodiversity on which PGRs thrives. Such a measure will be an important first step towards solving the problems of bio-piracy, and unrecompensed appropriation of traditional knowledge.

Michael Blakeney agrees with the need to provide a mechanism for compensating centres of origin for the value they have continued to add to society as a whole. According to him, "the traditional knowledge of indigenous peoples throughout the world has played an important role in identifying biological resources worthy of commercial exploitation (Blakeney, 2001)." However, like Kloppenburg, he concedes that it is difficult to quantify the economic value of "biological diversity conserved (improved) by farmers for agriculture."

Some commentators have argued against the commodification of traditional knowledge and practices for various reasons. It is their position that such development will offend some cultural values and spiritual significance of these resources (Battiste and Henderson, 2000). Arguments of this nature fail to recognize the property interests in traditional knowledge and state-occurring PGRs. Moreover, a formal extension of proprietary rights such as patent protection or a *sui generis* proprietary interest over these communally (indigenous) improved germplasm and the associated knowledge does not mean commodification. It merely acknowledges the following two key things. First, the economic values in these items and second, that in the biotechnology world, economic indices have overshadowed other means of measuring value. It is therefore for each source community with the help of its sovereign state to decide how to part with its PGRS and the traditional knowledge about them—either freely or on receipt of valuable consideration. However, without such recognition the demand for respect *simpliciter* of traditional or indigenous knowledge appears utopian.

BIOTECHNOLOGY INNOVATION AND THE BASIS FOR IPRS

Unlike intellectual property rights, the usual conception of property or property rights per se is in connection to tangible things. IPRs deal with ideas, which are treated as monopoly rights if they fulfill certain criteria. Depending on the nature of the idea and the manner of communication to the public, IPRs protection may be in the form of copyrights, patents, trademarks, industrial designs, or geographical indications.

Evenson, like some economists, argues that "patent and copyright protection were the chief means of providing incentives for invention" (Evenson, 1999). This position emphasizes the benefits of patents to innovators. Phillips and Khachatourians concur with Evenson when they stated, "ultimately, the challenge of examining innovation is in its quantification for its contributing value to rapidly evolving user needs and *significantly better return on investment*" (Phillips and Khachatourians, 2001). Similarly the description of Morck and Yeung (2001) on what "economists know, suspect and guess about the underlying determinants of the pace of innovators" buttresses this view. They envision an industrial democratic (free market) environment where market dynamics rather than "artificial distortions" (in the form of government incentives or subsidies) thrive.[22] The key to invention, they assert, is proactive investment in research and development. Under this arrangement, vital information is utilized for innovation purposes to maintain a competitive edge.[23] Without patent protection, information or ideas that give rise to innovation are susceptible to "free rider" problems (Morck and Yeung, 2001).

Free riding has been described as "situations where some individuals (albeit institutions and corporations also) "free ride" on the efforts of other individuals to provide either the good itself or the set of rules (and their monitoring and enforcement)" for sustaining a resource (Becker and Ostrom). Surprisingly, the free rider problem is not a major concern in the decision to deny IPR protection to traditional knowledge of the uses of plant germplasm. This being the case, are there inherent limitations or other convincing reasons why the incontrovertibly valuable indigenous knowledge about the uses of PGRs is excluded from patent protection? Why is the indigenous knowledge with respect to the uses of PGRs said to be in the public domain when a significant body of economic literature contends that such knowledge was never documented (Crucible II Group, 2000)? In contrast, the Crucible Group has confirmed that there exists proper documentation of traditional knowledge in local communities, albeit in a non-Western form. In one of their reports, the group asserts:

> Yet, scientific researchers (and Crucible members) such as Bo Bengtsson, Joachim Voss, Bernard Le Buanec and Louise Sperling—among others— have studied and assessed the contribution of farmers in Africa to the management of their genetic resources for food security and productivity. For example, they found that Ethiopian women would tabulate the yield results of their sorghum harvests on doorposts every year. They observed that women selected the highest-yielding, hardiest or otherwise most useful seeds from the field before men were allowed to harvest, and that these seeds would be "tested" in kitchen plots and even exchanged with neighbors for trials in differing soils. This is experimentation and documentation. The energetic exchange of seed between farming communities, common around the world, was an effort to access diverse research material to improve food security. Whereas short decades ago, conventional science sometimes described farmers' varieties as "Stone Age" or "primitive", the last few years have seen a shift in thinking and a much more realistic analysis of these varieties as ever-evolving and displaying differences from the seeds of the previous season.

The traditional knowledge about state-occurring PGRs is not static? Source communities have continued to improve it from one generation to another. It is therefore appropriate, except where they voluntarily decline, to compensate these communities for past-unrecompensed uses as well as for their present and ongoing knowledge on the uses of PGRs. After all, this is the same way Western scientists are compensated for their discoveries through the monopoly rights of intellectual property. Under this later development, the ability of scientists to tap into prior art is not questioned as long as there is evidence of improvement. This ability to rely on earlier discoveries is described in scientific communities as "standing on the shoulders of earlier scientists" (Scotchmer, 1991).

Understandably, under the Eurocentric test for patent protection, some traditional knowledge about germplasm relates to knowledge that may not strictly fit into the requirement of novelty. For instance, the knowledge may have been communicated to persons who were not originally part of its creation or development. But this communication often takes place within the communities and in consonance with their culture—founded on exchange and openness. Traditional knowledge exists first and foremost amongst local farmers and/or indigenous peoples, and within their communities. The free communication of traditional knowledge amongst these communities does not strictly jeopardize its essence as something new outside PGR source communities. Article 8(j) of the CBD recognizes this aspect of communal livelihood, inventions and practices of indigenous communities.[24]

Therefore, the argument for denying IPRs protection to PGRs appears inapplicable to the traditional knowledge of source communities about the various uses of germplasm. Although this knowledge is usually passed down from one generation to another, each generation adds to the existing knowledge in the same way that scientists utilize existing knowledge to innovate. If closely monitored, the various stages of generational development in traditional knowledge can be properly isolated for recognition. Propertization of traditional knowledge for the uses of PGRs would be a step towards meeting this objective. Considering the nature of property right as a social construct, this will be possible notwithstanding the cultural, spiritual, social and symbolic importance of such plants to source communities or centres of origin (Shiva, 1997; Rhoades and Nazarea, 1999). Until this proprietary imbalance is eliminated, the current international regime on biotechnology will remain a manipulation by the North, in favour of multinational corporations[25] involved in plant breeding, against the poor cultivars of the South (Wallerstein 1991, 2000, 2001).

CONCLUSION

This chapter analyses the basis for the exclusion together with the current dichotomy in the intellectual property rights' regime on the proprietary nature of PGRs and the associated traditional knowledge. It clarifies the nature of these resources against what is protectable under the existing intellectual property rights paradigm. Lastly, it challenges the basis of this exclusion under the guise of common heritage or their non-qualification for patent protection because they are already in the public domain. Unfortunately, the current IP regime remains a socio-political construct institutionalized and used by powerful states to protect and then exclude outsiders from products or knowledge originating from their territory. This is akin to Cox's description of the market as a socio-political construct driven by rational interactions between economic agents (Higgott 1994).

It is apparent that until issues surrounding the extension of patent protection to plant and animal life forms, genetic information and other items of discovery are resolved, developing countries will continue to feel marginalized at multilateral trade institutions such as the World Trade Organization (WTO) (Gibbs and Bromley, 1989).[26] Of utmost importance in this debate is the unequivocal request from the South for a redefinition of the international patent regime in order to recognize non-Western innovative practices. Developing countries are concerned over the ownership or proprietary control of their PGRs, especially with the continued notion, reference and/or treatment of these resources as common heritages of humankind by industrialized countries. It is only equitable that the common heritage notion over South occurring PGRs is discarded since their counterparts from the North's "elite" cultivars are private property. This is in addition to whether or not traditional knowledge of source communities and indigenous peoples concerning the uses of germplasm should be excluded from patent protection for lack of Western style documentation or publication notwithstanding the incontrovertible evidence that such knowledge belongs to a known and identifiable community. Indeed, diversity involves recognition of different property owning structures in various parts of the world. If the appropriate and enforceable property regime is established to protect PGRs and the related traditional knowledge, then communities that conserve and improve them will be properly compensated. While the CBD, although flawed in many respects, provides a starting point through its access and benefits sharing scheme. After all, without legal backing to prevent incidences of non-rivalry and non-excludability, "most discoveries and inventions exhibit public good attributes" (Moschini, 2003).

The controversy over the proprietary status of PGRs and traditional knowledge about their uses, together with the nature of the benefit sharing and technology transfer arrangement in the CBD, resembles a class struggle in the Gramscian sense. If sustained, the controversy may lead to the WTO losing its leading role as the international institution driving free trade. Developing countries, like the oppressed class in any society, organization or arrangement appear to understand that their interests may not after all be in the WTO. Their walkout from the 2003 WTO Ministerial Conference in Cancun, Mexico when they did not receive the anticipated concession on key demands is a confirmation of this awareness. However, the Cancun approach suggests a lack of the structural power,[27] influence and bargaining strength required to obtain favourable conditions at the WTO negotiating table. Such a structural power will be required to change the commons notion erroneously extended by the North to PGRs and the associated traditional knowledge.

NOTES

[1] In this chapter, "South" is used in reference to "developing" as well as the "less developed" countries of Asia (excluding Japan), Africa, Latin and South America. They are also collectively referred to as "centers of origin" or "source communities." Besides low per capita income, these countries with the exception of China, South Africa, India, Brazil, and Argentina have little or non-existent investment or research and development in agricultural biotechnology.

[2] The word "North" or "Western," as used in this chapter, refers to "developed" countries of Europe, North America and Japan. These countries have commendable per capita income in addition to extensive investment in both medical and agricultural biotechnology. Both terms have the same meaning or connotation with "West" which appears in some of the literatures referenced.

[3] Although shielded from private ownership, common resources such as the high seas and the airspace are available to facilitate the economic and personal inter-relationship of humanity. These resources are still subject to regulations primarily for public good and interests (see Gray, 1991; Rose, 1986).

[4] For instance, on moral non-excludable resources, Gray stated: "[T]he notion of moral non-excludability derives from the fact that there are certain resources which are simply perceived to be central or intrinsic to constructive human co-existence that it would be severely anti-social that these resources should be removed from the commons." (Gray, 1991).

[5] In his definition of property, Pufendorf stated, "[o]wnership is a right, which what one may call the substance of a thing belongs to someone in such a way that it does not belong in its entirety to anyone else in the same manner."

[6] Locke's property theory is based on what is commonly known as the Lockean "labour theory."

[7] Hegel describes property as an "expression of self" and his justification for propertization is known as the "personality theory."

[8] Such as the Japanese *Tale of Genji*.

[9] (1921) A.C. 399 (Appeal Case Report).

[10] Consequently, Hardin could have talked of the "Tragedy of Open Access Resources" as against the "Tragedy of the Commons." For a criticism of Hardin's error in this regard, see Peters (1987) and Townsend and Wilson (1987).

[11] Please note that unlike *res nullius* things that fall within *res communis* may be owned. As Engle explained, "*res communis* can be owned, and are owned, by the state, though a state may permit anyone to appropriate" it as was done by the Roman state. (Engle, 2004, Shaw, 1997). Shaw argues that *res communis* is incapable of sovereign control or private ownership. He equates *res communis* with the common heritage concept, to wit, something that "belongs to no one but may be used by all."

[12] See for instance Part XI of the 1982 United Nations Convention on the Law of the Sea (UNCLOS) to the effect that "the Area [i.e. the seabed and ocean floor beyond the limits of national jurisdiction] and its resources are the common heritage of mankind (sic)." See Art. 137, UNCLOS, 1982.

[13] Treaty on Principles Governing the Activities of States in the Exploration and Use of the Outer Space, including the Moon and other Celestial Bodies, 610 UNTS 205.

[14] See the WIPO Report on Fact-finding Missions on Intellectual Property and Traditional Knowledge (1998–1999).

[15] The UN Convention on Biological Diversity (CBD) recognizes this responsibility and attempts to institutionalize a distributive wealth maximization mechanism to compensate source communities through their states for unrecompensed benefits obtained from uses of their PGRs and the traditional knowledge thereof.

[16] As Knorr (1973) rightly stated, "[W]hile the economic penetration of weak states has frequently preceded their formal incorporation into the empire of the strong state, thus entailing a complete loss of sovereignty, economic penetration is also one way for one state to gain prolonged domination—economic and political—over a weaker state which retains its formal sovereignty."

[17] UPOV was established pursuant to the International Convention for the Protection of New Varieties of Plants, 1961. The Convention was adopted in Paris, France and last revised in 1991. For a quick overview of this Convention, see Ghijsen, (1998).

[18] Nevertheless, following persistent pressures by developing countries and Non-Governmental Organizations (NGOs), Farmers' Rights over seeds and crop plants were recently re-affirmed (November 2001) by the newly adopted International Treaty on Plant Genetic Resources for the Food and Agriculture Organization. Despite this treaty's apparent conflict with UPOV, it is now operative following its coming into force on 29 June 2004.

[19] These are novelty, inventiveness and commercial viability. See Article 17.1 of the TRIPs Agreement.

[20] "TK" as used by Dutfield is an acronym for Traditional Knowledge.

[21] This is the "International Treaty on Plant Genetic Resources for Food and Agriculture," adopted by the FAO Conference on 3 November 2001, and came into effect on 29 June 2004.

[22] But are patents too not "artificial distortions" of sort? Cf. with other possible systems of rewarding innovators such as outright payment by government. This, as argued by some commentators, is a better way to compensate innovators for their ingenuity rather than through the concept of patent protection. (Shavell and van Ypersele, 1998).

[23] It is an age in which information rules.

[24] Cf. William Lesser (2000) who claims that the contribution of indigenous people through traditional knowledge to biotechnology are "very intangible" and that "compensating this group for its contributions, and providing incentives for continued conservation [improvement] of biodiversity, will require additional consideration, possibly a new form of formalized rights".

[25] The capitalist inclinations of the North, which Wallerstein described as "a system that has imperative need to expand—expand in terms of total production, expand geographically—in order to sustain its prime objective, *the endless accumulation of capital*," is currently projected through the MNCs on the issue of biotechnology.

[26] Similarly, Kevin Gray suggests that "[m]uch of our false thinking about property stems from the residual perception that "property" is itself a thing or resource rather than a legally endorsed concentration of power over things or resources." He then submits, "the limits of "property" are fixed, not by the "thinglikeness" of particular resources but by the physical, legal and moral criteria of excludability" (see Gray, 1991).

[27] Strange identified four main structural powers, namely the Knowledge Structure, the Security Structure, the Production Structure, and the Financial Structure (see Strange, 1994).

REFERENCES

Antarctic Treaty, concluded in Washington DC, on 1 December 1959, entered into force on 23 June 1961, reprinted in 40 UNTS 71

Aoki K (1998) Nationalism, anticommons property, and Biopiracy in the (Not-So-Brave) new world order of international intellectual property Protection 6 Ind. J. Global Leg. Stud. 11. Retrieved from: http://www.law.uoregon.edu/faculty/kaoki/site/articles/notsobrave.pdf

Charter of Economic Rights and Duties of States (1974) Article 3

Battiste M, Henderson JY (2000) Protecting indigenous knowledge and heritage: a global challenge. Purich, Saskatoon

Becker DC, Elinor O (1995) Human ecology and resource sustainability: the importance of institutional diversity. *Annual review of ecology and systematics, vol* 26. pp 113–133

Blakeney M (2001) Intellectual property aspects of traditional agricultural knowledge. In: Drahos P, Blakeney M (eds) IP Biodiversity and agriculture: regulating the biosphere. Sweet and Maxwell, London

Brownlie I (1998) Principles of public international law, 5th edn. Oxford University Press, Oxford

Busch L, Lacy WB, Burkhardt J, Lacy LR (1991) Plants, power and profit: social, economic, and ethical consequences of the new Biotechnologies. Basil Blackwell, Cambridge MA and Oxford

Carr CL (ed) (1994) The political writings of Samuel Pufendorf. trans. by Seidler MJ. Oxford University Press Oxford

Convention on Biological Diversity. Article 2(2). available online: http://www.biodiv.org/convention/articles.asp checked 1993

Crucible II Group (2000) Seedling solutions: policy options for genetic resources (people, plants, and patents revisited), vol 1. IDRC & IPGRI, online version, Ottawa & Rome

Dietz T, Elinor O, Stern PC (2003) The struggle to govern the commons, Science,12/12/2003, vol 302 Issue 5652. pp 1907–1912

Drahos P (1996) A philosophy of intellectual property. Dartmouth, Aldershot

Dutfield G (1999) Protecting and revitalising traditional ecological knowledge: intellectual property rights and community knowledge databases in India. In: Blakeney M (ed) Intellectual property aspects of Ethnobiology. Sweet & Maxwell, London

Dutfield G (2001a) Indigenous peoples, Bioprospecting, and the TRIPS Agreement: threats and opportunities. In: Drahos P, Blakeney M (eds) IP in biodiversity and agriculture: regulating the biosphere. Sweet and Maxwell, London

Dutfield G (2001b) TRIPS-related aspects of traditional knowledge. Case Western Reserve Journal of International Law 33(2) JIL 233-275

Engle E (2004) Economic theory of Law and the public domain: when is piracy economically desirable. available online: http://lexnet.bravepages.com/media1.html

Epstein RA (1994) On the optimal mix of private and common property 11 social philosophy and policy 17. reprinted in Epstein RA (ed) Private and common property. Garland, New York & London

Epstein RA (2000) Possession as the root of title. In: Epstein RA (ed) Private and common property.Garland, New York & London

Evenson RE (1999) Intellectual property rights, access to plant germplasm, and crop production scenarios in 2020" (1999) 39 Crop Science 1630

Eyzaguirre P, Evan D (2003) The impacts of collective action and property rights on plant genetic resources – being a paper prepared for the CAPRi-IPGRI workshop on Property rights, collective action and local conservation of genetic resources (in Rome), available online: http://www.capri.cgiar.org/pdf/GReyzaguirre.pdf

Federal Court Trial (FCT) 256 (2001) available online: http://decisions.fct-cf.gc.ca/fct/2001/2001fct256.html The Supreme Court of Canada's judgment on this landmark case is reported in *Monsanto Canadav. Schmeiser*[2004] 1 S.C.R. 902.

Food and Agriculture Organization, International Undertaking on Plant Genetic Resources. 1983 Available online: http://www.fao.org/ag/cgrfa/IU.htm

Fowler C, Smale M, Gaiji S (2001) Unequal exchange? Recent transfers of the agricultural resources and their implications for developing countries. Development Policy Review 19(2):181–204

GATT/WTO (1995) Agreement on Trade Related Aspects of Intellectual Property Rights (TRIPs), which came into effect on January 1, 1995.

Gertler ME (1998) Biotechnology and social issues in rural agricultural communities: identifying the issues. In: Hardy RWF, Segelken JB, Voionmaa M (eds) Resource management in challenged environment. NABC, New York

Ghijsen H (1998) Plant variety protection in a developing and demanding world. Biotechnology and Development Monitor 36:2–5

Gibbs CJN, Bromley DW (1989) Institutional arrangements for management of rural resources: common property regimes. In: Berkes F (ed) Common property resources: ecology and community-based sustainable development. Belhaven Press, London

Goldstein D (1988) Molecular biology and the protection of germplasm: a matter of national security. In: Kloppenburg JR (ed) Seeds and sovereignty: the use and control of plant genetic resources. Duke University Press, Durham

Gray K (1991) Property in thin air. Cambridge L J 50:253–256

Greenpeace Canada (2002) New report exposes multiple threats if GE wheat approved available online: http://www.greenpeace.ca/e/index.php

Grotius H (1925) De Jure Belli Ac Pacis Libre Tres, trans. by Kelsey FW (1964). Wildy, New York & Oceania

Hardin G (1993) The tragedy of the commons. In: Daly HE, Townsend KN (eds) Valuing the earth: economics, ecology, and ethics. The MIT Press, Cambridge, MA

Harris JW (2001) Property and justice. Oxford University Press, Oxford & New York

Heller M (1998) Tragedy of the anticommons: property in the transition from Marx to markets. Harv L Rev 11:621

Higgott R (1994) International political economy. In: Groom AJR, Light M (eds) Contemporary international relations: a guide to theory. Pinter Publishers, London & New York

Hughes J (1988) The philosophy of intellectual property. Geo L J 77:287

International Convention for the Protection of New Varieties of Plants (UPOV Convention) (1978) as amended in March 1991, and 1996, respectively. Online at: http://www/upov.int

Kelly D, Michelman F (1980) Are property and contract efficient? Hofstra L Rev 8:711

Kloppenburg JR, Kleinman DL (1988) Seeds of controversy: National property versus common heritage. In: Kloppenburg JR (ed) Seeds and sovereignty: the use and control of genetic resources. Duke University Press, Durham & London

KloppenburgJR (1988) First the seed: the political economy of plant biotechnology, 1492–2000. Cambridge University Press, Cambridge & New York

Kloppenburg J, Kleinman DL (1987) The plant germplasm controversy. BioScience 37(3):190–198

Knorr K (1973) Power and wealth: the political economy of international power. Basic Books, New York

Krasner SD (2000) State power and the structure of international trade. In: Friedman JA, Lake DA (eds) International political economy: perspectives on global power and wealth. Bedford/St. Martin's, Boston & New York

Laslett P (ed) (1960) Two treatises of government: a critical edition. Cambridge University Press, Cambridge

Lesser W (2000) Intellectual property rights under the convention on biological diversity. In: Santiello V, Evenson RE, Zilberman D, Carlson GA (eds) Agriculture and intellectual property rights: economic, institutional and implementation issues in Biotechnology. CABI, London

Locke J (2001) Two treatises of government. In: Pecora VP (ed) Nations and identities: classic readings. Blackwell, Malden, MA

Low A (2001) The third revolution: plant genetic resources in developing countries and China: global village or global pillage? In: Kinsler J et al (eds) International trade and business law annual. Cavendish, Newport, New South Wales

Mgbeoji I (2002) Patents and plant resources-related knowledge: towards a regime of communal patents for plant resources-related knowledge. In: Islam N et al. (eds) Environmental law in developing countries: selected issues. IUCN, Bonn, Germany

Mgbeoji I (2001) Patents and traditional knowledge of the uses of plants: Is a communal patent regime part of the solution to the scourge of bio piracy? Ind J Global Legal Stud 9:163–186

Morck R, Yeung B (2001) The economic determinants of innovation, Industry Canada Research Publications No.25

Moschini G (2003) Intellectual property rights and the World Trade Organization: retrospect and prospects. Working Paper 03-WP 334, also available online: CARD website: www.card.iastate.edu

Mossoff A (2003) What is property? Putting the pieces back together. Ariz L Rev 45:371

Murphy SD (2001) Biotechnology and international Law. Harv Int'l L J 42:47

Onwuekwe CB (2004) The commons concept and intellectual property rights regime: Whither plant genetic resources and traditional knowledge? Pierce 2(1) Law Review 65–90

Ostrom, E.(2004) Understanding Collective Action. In Ruth N. Meinzen - Dick and Monica Di Gregorio, *Collective Action and Property Rights for Sustainable Development*. brief 2.2020 Four service, No.11, Washington, D.C.: International Food Policy Research Institute.

Ostrom, E.(1999) Coping with Tragedies of the commons. *Annual Review of Political Science* 2,pp. 493–535

Peters PE (1987) Embedded systems and rooted models: the grazing lands of botswana and the commons Debate. In: MaCay BJ, Acheson JM (eds) The Question of the Commons: the Culture and Ecology of Communal Resources. The University of Arizona Press, Tucson, AZ

Phillips P (2001) Will biotechnology feed the sorld's hungry? Int'l J 56(4):665–677

Phillips PWB, Khachatourians GG (2001) Approaches to and measurement of innovation. In: Phillips PWB, Khachatourians GG (eds) The Biotechnology revolution in global agriculture: invention, innovation and investment in the Canola sector. CABI, Oxon & New York

Pistorius R, van Wijk J (1999) The exploitation of plant genetic information: political strategies in crop development. CABI, Oxon & New York

Posey DA, Dutfield G (1996) Beyond intellectual property: toward traditional resource rights for indigenous peoples and local communities. IDRC, Ottawa

P.L. (Public Law) 91-577. 1970. Plant Variety Protection Act, 84 Stat. 1542–1559

RAFI (1996) Enclosures of the mind: intellectual monopolies. IDRC, Ottawa

Rhoades RE, Nazarea VD (1999) Local management of biodiversity in traditional agroecosystems. In: Collins WW, Qualset CO (eds) Biodiversity in agroecosystems. CRS Press, London & New York

Rose CM (1986) The comedy of the commons: customs, commerce, and inherently public property. U Chi L Rev 53:711

Scotchmer S (1991) Standing on the shoulders of giants: cumulative research and the patent law. J Econ Perspect 5:29

Sedjo RA (1988) Property rights and the protection of plant genetic resources. In: Kloppenburg JR (ed) Seeds and sovereignty: the use and control of plant genetic resources. Duke University Press, Durham & London

Shavell S, van Ypersele T (1998) Rewards versus intellectual property rights available online: http://www.law.harvard.edu/programs/olin_center/papers/pdf/246.pdf

Shaw MN (1997) International Law, 4th edn. Cambridge University Press, Cambridge & New York

Shiva V (1993) Monocultures of the mind: perspectives on biodiversity and Biotechnology. Zed Books, New Jersey

Shiva V (1997) Biopiracy: the plunder of nature and knowledge. Between the Lines, Toronto

Stenson AA, Gray TS (1999) The politics of genetic resource control. St. Martins Press, New York

Strange S (1994) States and markets, 2nd edn. Pinter, London & New York

Townsend R, Wilson J (1987) An economic view of the tragedy of the commons. In: McCay B, Acheson J (eds) The questions of the commons–the culture and ecology of communal resources. University of Arizona Press, USA

Trade Related Aspects of Intellectual Property Rights (TRIPs). TRIPS Article 27.1 of the Agreement. Available online: http://www.wto.org/english/tratop_e/trips_e/t_agm0_e.htm

Tully J (1980) A discourse on property: John Locke and his adversaries. Cambridge University Press Cambridge & New York

UN General Assembly, Twenty-Eighth Session, Resolution 3129 XXVIII (1973) Cooperation in the field of the environment concerning natural resources shared by two or more States. Available online: http://www.un.org/documents/ga/res/28/ares28.htm

UNCLOS (with Annex V), Art. 137, concluded at Montego Bay, 10 December 1982, and entered into force on 16 November 1994. UN Doc. A/CONF.62/122; reprinted in (1982) 21 I.L.M. 1261

United Nations Conference on Environment and Development (UNCED) (1992) Rio de Janeiro

Utuama AA (1991) Customary Law and land use act. In: Osinbajo Y, Kalu AU (eds) Towards a restatement of Nigerian customary Laws. Federal Ministry of Justice, Lagos

Wallerstein I (1991) Geopolitics and geoculture: essays on the changing world–system. Cambridge University Press, Cambridge & New York

Wallerstein I (2000) The modern world–system. In: Garner R (ed) Social theory: continuity and confrontation –A reader. Broadview Press, Ontario

Wallerstein I (2001) The end of the world as we know it: social science for the twenty-first century. University of Minnesota Press Minneapolis

World Intellectual Property Office (WIPO) (1999) Report on fact-finding missions on intellectual property and traditional knowledge. available online: http://www.wipo.int/tk/en/tk/ffm/report/final/index.html

Zilberman D, Yarkin C, Heiman A (1998) Institutional change and Biotechnology in agriculture: implications for developing countries. In: Smale M (ed) Farmers, gene banks and crop breeding: economic analysis of diversity in wheat, maize, and rice. Kluwer Academic, Boston & London

PETER W.B. PHILLIPS

CHAPTER 3

FARMERS' PRIVILEGE AND PATENTED SEEDS

INTRODUCTION

The global agri-food industry has reoriented in the past few decades around technological change and innovation. Both farmers and the rest of the agri-food supply chain have recognized that the long-term threat to their livelihoods is other local and regional demand for land, labor and capital. Ultimately, the agri-food sector must deliver productivity gains at least equal to other domestic sectors, or mobile land, labour and capital resources will be bid away. This has precipitated a global debate about the appropriate balance between the rights and obligations of farmers on the one hand and investors or inventors on the other.

In the not so distant past, most of the new plant varieties in most open pollinated plants were developed by publicly funded research programs or institutes and were commercialized on a concessionary basis (often given to farmers at nominal or no charge), with seed production, industrial inputs, production, handling, processing and retailing all being handled through arms-length market transactions. The scale and complexity of using and commercializing products of biotechnology, a highly globalized science, has precipitated collaborations between traditional competitors and between public and private research organizations, which has required new institutions in support of development. In response, governments have encouraged the private sector to take the lead in the search for new technologies and products with new monopoly intellectual property rights (both patents and plant breeders' rights) and by new or different forms of government subsidy and support. New private rights to inventions related to seeds and the resulting complementarities between privately developed germplasm and industrial chemicals have caused new integrated relationships to emerge among biotechnology seed developers, chemical companies and farmers. One implication is that the results of the research—both new technologies and new plant varieties—have been exploited in narrow monopolistic situations, which on the face of it has the potential to reduce the social benefits of these investments.

This change from a publicly led research, development and commercialization system to an increasingly privately directed system has created controversy. A recent Canadian court case involving Saskatchewan farmer Mr. Percy Schmeiser and Monsanto Canada Inc. has highlighted the trade-offs inherent in private ownership of new innovations. Monsanto introduced a transgenic, Roundup Ready (RR) canola variety in 1995 in Western Canada and used a wide range of mechanisms to protect its investment in intellectual property, including: patents on the processes used to develop the variety and on the proprietary RR gene itself; plant breeder's rights on the variety; and a private contract (called a Technology Use Agreement, or TUA) on the resulting commercial seed sale. All of the technologies and properties were also protected by trademarks in the marketing chain.

This case highlighted the tradeoffs inherent in the new research strategy in Canada and more broadly. While plant breeders' rights around the world explicitly allow farmers to save and reuse seed for sowing their own crops (but not to multiply the seed and sell it to others for sowing), patents and private contractual arrangements (in this case the TUA) both specifically prohibit saving seeds for any purpose. In Mr. Schmeiser's case, he acquired seed through an undetermined way, and hence was not party to the contract. When Monsanto challenged his possession of seed that contained a high percentage of the RR trait, Mr. Schmeiser responded that he was entitled to save and reuse seed as provided through "farmers' privilege" in the Canadian Plant Breeders' Rights Act. Monsanto responded with a civil action against Mr. Schmeiser under the Patent Act, claiming, amongst other things, that he did not have an appropriate license to use their patented technology.

Throughout the trial in 2000–1, and in the subsequent appeals to the Federal Court of Appeal and the Supreme Court of Canada, a wide range of issues were raised. The respective pleadings of Monsanto and Mr. Schmeiser raised a wide range of questions about whether it was appropriate to use biotechnology at all, the impact of the increasing dependence on private rather than public breeding, the rights of an accused to due process, the impact of new technologies and patent enforcement on social cohesion in rural communities and the environmental, health and economic impacts of the technology on ecosystems, consumers and farmers. Throughout the entire proceedings, the issue of farmer's privilege dominated—should patents be able to be applied either to parts of or to an entire multicellular organism, and if so, should farmers have the privilege of saving that seed, whether patented or not, to resow in subsequent years without further royalty payments?

This chapter examines the role of controversies and assumptions underlying the differing perspectives on farmers' privilege and commercial seed production, offering a review of the legal basis for farmers' privilege in domestic and international law and some observations on the challenges to the current system.

EVOLUTION OF IPRS FOR AGRI-FOOD TECHNOLOGIES AND PRODUCTS

There have been different forms of intellectual property protection on plant varieties for thousands of years. There is some evidence that the Egyptian, Greek and Roman empires attempted to control access to particular crops, seed varieties or livestock (Smyth et al., 2002). More recently, the imperial government in India in the 1700s attempted to control access to rice seed (by threatening capital punishment for anyone that exported seed for breeding) and many of the colonial trading houses endeavored to control access to the seeds that were used in their plantations. By the 1930s, plant breeders had developed some commercially viable hybrids, as in corn, which by their very nature discourage farmers from sowing saved seed. Even though some of these IPR mechanisms worked at some points in time, they tended to be effective only in a few discrete product markets, especially where there were some natural or biological barriers to transferring germplasm.

The earliest formalized system of IPRs was the medieval guilds, where apprentices would indenture themselves for years in order to learn the trade secrets, craft and arts of the guild. The covenant would have civil penalties, up to and including execution. These systems were very effective at concentrating power in the hands of members of the guilds (creating great wealth along the way), as well as inculcating quality assurance and research into the production system. Venice is often credited with having the first formal patent law, dating back to 1474. Over the following centuries this model was adapted and adopted across much of Europe. The resulting patent system provided formal rights to use and to exclude others from using "inventions" that were novel, useful and non-obvious.

The main impetus for formalizing private rights to invention in the agri-food world came from the United States. As the single largest source of new inventions and the single largest market for those inventions, the US plays a vital role in almost all innovation-based sectors. Although the US began in 1776 with little or no patent or copyright protection (patents were formalized in 1790 through the US Patent Act), it now is viewed as the prime mover and supporter for private rights to inventions and innovations. In 1935, the US government passed an amendment to the US Patent Act providing for "plant patents" for germplasm that was asexually reproduced (e.g. by cuttings or grafting, hybridization, spore or mutation), which supported increased investment in research and development in various hybrid crops such as corn, soft-fruits and trees. Meanwhile in Europe, during the 1940s and 1950s a number of states developed plant breeders' rights that accommodated and protected varieties of plants that reproduced sexually. In 1961, a number of European countries, along with a few other states, negotiated the International Convention for the Protection of New Varieties of Plants (known as UPOV after its French acronym), which provided for an international recognition of breeders' rights in signatory states. Essentially, signatories agreed to extend plant variety protection reciprocally to varieties developed in other member countries. This agreement was revised in 1978 and 1991. As of January 2004, 54 countries had signed one of the three agreements. In 1970 the US developed the Plant Variety Protection Act (PVPA) and acceded in 1981 to the UPOV, thereby granting property rights to sexually reproduced varieties, such as open or self-pollinated varieties. As discussed in more detail below, these new rights (both in the PVPA of 1970 and in UPOV) were limited by two exemptions. Farmers were deemed to have the privilege to save and reuse seed from protected crops (but not to resell it to other farmers for sowing) and researchers were allowed to use protected germplasm for research and development purposes.

While the plant variety protection system was evolving, there was a merging of the sciences, such that patents began to be approved for an increasing array of technologies and components of plant varieties. On the technology front, in 1973 the US Patent Office granted Cohen and Boyer a utility patent on gene splicing technology, starting the race to privatize agronomic research. Under US patent law (and most other comparable national systems), patents are available for "anything under the sun" that is the product of human ingenuity, provided it is a "new, useful and non-obvious process, machine, article of manufacture, or composition of matter, or any new and useful improvement thereof" (US PTO, 2004). Every patent offers exclusive rights to use and exclude others from using the patent, for a period of 20 years beginning with the date of filing. Since 1973, virtually all of the main technologies required for genetic manipulation of a plant or animal have been patented in the US (see Phillips and Khachatourians, 2001; Nottenburg et al., 2003). To be eligible for patent, inventions must not have been known or used by others in the US, have been patented or described in a printed publication in the US or any foreign country or have been in public use or on sale in the US for more than one year prior to the date of application. Inventions that are novel, useful and non-obvious are then eligible for protection. Although US patent protection is not automatically accepted in other countries, most of the key technologies have been patented in prime markets (for details, see the FTO analyses undertaken by CAMBIA in recent years).

In the 1980s, a number of landmark rulings related to patenting living organisms opened the floodgates. In 1980 the US Supreme Court ruled in *Diamond* v. *Chakrabarty* that the US patent law provides for patenting life-forms. The first patent on a life form was for a genetically modified oil-eating bacterium. In 1985 the first patent for a living plant was issued. Since then a number of plants have been patented. Plant patents provide additional protection to inventors and innovators in addition to those in UPOV in that plant patents do not provide for either a research exemption or farmers' privilege.

These moves precipitated a response in most other major countries with indigenous research and development (Table 1). Europe, Japan and Australia, for example, all developed comparable plant variety systems (consistent with the latest UPOV agreement reached in 1991) and, through judicial review, extended patents to single and multicellular organisms as in the US. Many developing countries, in contrast, have been slower to develop indigenous IPR systems. With the incorporation of the Agreement on Trade-Related Intellectual Property Rights (TRIPs) in the WTO Agreement of 1995, all member states of the WTO (numbering 149 in early 2005) are now required to operate or develop intellectual property systems (patents, plant breeders rights or other *sui generis* systems) to protect all manner of inventions, including technologies, genes and plant and animal varieties. Furthermore, member states must provide an effective system of enforcement and adjudication of rights. Highly developed countries (e.g. US and EU) had to conform by 1996, about 70 leading developing nations (e.g. Argentina) and transition economies had until 2000 to conform, 30 least-developed economies (e.g. Bangladesh) have until 2006 to comply and new member states (e.g. China) have 1 year after accession to comply.

Canada has been a bit of an outlier relative to other countries with significant investments and interests in agricultural biotechnology and agri-food R&D

TABLE 1. Key policies affecting commercialization of intellectual property in competing countries, 2004

	Utility patents	Patents on plants	UPOV (date joined & current agreement)	Conform with TRIPS	Private contracts	Legal enforcement
Argentina	Weak	No	1994; UPOV 1978	?	Yes	Weak (no effective enforcement/penalties)
Australia	Yes	?	1989; UPOV 1991	Yes	Yes	Moderate (limited punitive damages)
Canada	Yes	No	1991; UPOV 1978	Yes	Yes	Moderate (limited punitive damages)
China	Yes	?	1999; UPOV 1978	Yes	Limited	Limited (no effective enforcement/penalties)
EU (17 member states)	Yes	Yes	Earliest, 1968; UPOV 1991	Yes	Yes	Moderate (limited punitive damages)
India	Weak	No	Not a member	?	Yes	Weak (no effective enforcement/penalties)
Japan	Yes	?	1982; UPOV 1991	Yes	Yes	Moderate
US	Yes	Yes	1981; UPOV 1991	Yes	Yes	Strong due to punitive damages

Sources: WIPO; UPOV; WTO; Authors estimations.

(Table 1). All major industrial countries now conform to both TRIPs and UPOV 1991, which requires protection be provided, allowing innovators to use patents for all technologies and constructs and either PBRs or patents for plant varieties. Canada, in contrast, acceded only in 1991 to UPOV 1978, which does not provide for patents on plant varieties. Canada's Plant Breeders Rights Act provides 18 years of protection for new varieties and allows for both farmers' and research exemptions. Meanwhile, the Canadian patent system accommodated some of the demands for patents, extending utility patents as requested for new biotechnologies, for gene constructs and to single cellular organisms. Two recent court decisions have effectively extended patent control to plants. While the Canadian Intellectual Property Office has refused to grant patents for multicellular living organisms (supported by a Supreme Court 5/4 split decision in the Harvard Oncomouse case), the Supreme Court ruling in 2004 on *Monsanto* v. *Schmeiser* (once again a 5/4 spit decision) concluded that while Monsanto did not have the right to patent the entire RR canola plant, it did have the right to exclude any others from using their plants without a valid license. The Court ruled that because the RR variety in question had the RR gene cassette expressed in every cell of the plant, Monsanto should have under the Patent Act the right to exclude others from using their technology. Through this judgment, the Court would appear to have effectively extended patents on multicellular organisms (CIPP, 2004). The issue may arise again if new GM products are introduced where proprietary components are only expressed in parts of the organism, but that may be a while in the future.

Effectively, and as shown in Table 1 above, what has evolved is a complex web of IPR mechanisms that allow innovators to arrange a dense web of interlinked and overlapping property rights protections that mutually support each other. As a result, even if one system proves to offer weak protection in some circumstances, there are other options available.

THE HISTORY OF FARMER'S PRIVILEGE

The history of a formalized farmers' privilege is relatively recent. While the practice of farmers sowing crops with saved seed has been a common practice throughout millennia, the prerogative for farmers would appear to be more the result of the impracticality and inappropriateness of enforcing any private rights, rather than an overriding principle or right. In earlier times, innovation in agriculture and food production was often not the result of planned investment by profit making enterprises. Rather, it was either the result of serendipity or due to public investment in breeding programs at public universities and public labs. Given that the motivation was usually to support farmers, most of these inventions were freely distributed to any farmer who would take and use them. For example, in Canada, Marquis wheat and low erucic-acid, low-glucosinolate rapeseed (trademarked as canola) were the products of public investment and research and given to farmers free of charge. This system worked well as long as public investments continued and farmers were the main beneficiary of the innovations.

As science advanced and markets developed, the commercial imperative began to emerge. Many farmers and seed companies began to adapt new varieties to specific markets, adding incremental value to the seed that was only realizable in specific circumstances. As much as possible, these innovators attempted to capture as large a share of the new value as possible but a *de facto* farmers' privilege existed in that it was largely impracticable in most jurisdictions for small seed developers to seek out and attempt to recover benefits generated by farmer-saved seed.

The first even semi-formalized acknowledgement of farmers' privilege occurred in the first plant breeders' rights programs in a few European countries. In those cases, breeders' rights were specified as relating to commercial sale of seeds, remaining mute on what farmers might do with seed on farm (except to prohibit them from reselling it for seed). This form of undefined farmers' privilege then was incorporated into the International Convention for the Protection of New Varieties of Plants (UPOV) of December 2, 1961. Article 5 [Rights Protected; Scope of Protection] of the agreement stipulated that "(1) The effect of the right granted to the breeder of a new plant variety or his successor in title is that his prior authorization shall be required for the production, for purposes of commercial marketing, of the reproductive or vegetative propagating material, as such, of the new variety, and for the offering for sale or marketing of such material. Vegetative propagating material shall be deemed to include whole plants." By absence of what is not mentioned, farmers who simply sowed seeds saved from previous crops and then sold the resulting crop for food (and not seed) were not infringing the rights of PBR holders.

This informal "privilege" remained in force through the 1978 UPOV Agreement but in 1979 debates began in the Food and Agriculture Organization about the "asymmetric benefits derived by the donors of germplasm and the donors of technology." The FAO concluded that commercial varieties were usually the product of applying breeders' technologies to farmers' germplasm and, while the breeders were able to generate returns through PBRs or other property mechanisms, farmers were not compensated. The debates ultimately led to a series of FAO resolutions (4/89, 5/89 and 3/91) which formally recognized the concept of farmers rights as a "basis of a formal recognition and reward system, intended to encourage and enhance the continued role of farmers and rural communities in the conservation and use of plant genetic resources." (FAO CPGR-Ex1/94/5, September 1994). The logic was that farmers' privilege was needed to balance "the rights of traditional breeders and of plant breeders, while allowing the farmers to benefit, in some way, from the value that they have creatively contributed ... recogniz[ing] the role of farmers as custodians of biodiversity and ... to call attention to the need to preserve practices that are essential for a sustainable agriculture." These debates and resolutions led to two outcomes (www.southcentre.org). Within the policy community, it initiated discussions through the FAO, the Agenda 21 process and the Convention on Biological Diversity (CBD) to revise the International Undertaking on Plant Genetic Resources (IUPGR), ultimately leading to the International Treaty on Plant Genetic Resources for Food and Agriculture, adopted 3 November 2001. This treaty formalized management of the CGIAR seed banks as "common heritage of humankind."

Meanwhile, this debate within the FAO community helped to inform a renegotiation and revision to the UPOV agreement. Article 5 of the 1991 UPOV agreement (UPOV, 1991) further elaborated the rights of breeders and farmer, stating that:

(1) Subject to Articles 15 and 16, the following acts in respect of the propagating material of the protected variety shall require the authorization of the breeder:
    (i) production or reproduction (multiplication),
    (ii) conditioning for the purpose of propagation,
    (iii) offering for sale,
    (iv) selling or other marketing,
    (v) exporting,
    (vi) importing,
    (vii) stocking for any of the purposes mentioned in (i) to (vi), above.

Article 15 goes on to formalize the exceptions to the breeder's right, stating
(1) the breeder's right shall not extend to
    (i) acts done privately and for non-commercial purposes,
    (ii) acts done for experimental purposes and

(iii) acts done for the purpose of breeding other varieties, and, except where the provisions of Article 14(5) apply, acts referred to in Article 14(1) to (4) in respect of such other varieties.

(2) Notwithstanding Article 14, each Contracting Party may, within reasonable limits and subject to the safeguarding of the legitimate interests of the breeder, restrict the breeder's right in relation to any variety in order to permit farmers to use for propagating purposes, on their own holdings, the product of the harvest which they have obtained by planting, on their own holdings, the protected variety or a variety covered by Article 14(5)(*a*)(i) or (ii).

In one way or another, all of the 54 current member states of the UPOV (1961, 1978 or 1991) have incorporated the concept of farmers' privilege into their domestic acts. What is less clear is how this privilege is affected by any of the other intellectual property mechanisms also used in these countries (discussed in Chapter 1). Quite clearly there is no provision within any patent act for users privilege (except via compulsory licensing, which is increasingly restricted by international agreement). Trademarks similarly are not mutable (not even through compulsory licensing). Private contracts can either provide privilege or not—that is one of the terms that would be negotiated in the context of the specific relationship. Strategic options, such as bundling, co-marketing and co-production are similarly open to privilege. The one mechanism that is unambiguously not open to privilege is the array of biological control mechanisms, which limit germination of progeny seed (e.g. hybrids, seedless fruit and GURTs). Effectively, we have a mixed system of privilege that is often more respected in theory than in practice.

THE BATTLE OVER FARMERS' PRIVILEGE

The patent case involving the multinational agro-chemical giant Monsanto and a Canadian farmer, Mr. Percy Schmeiser, has brought the issue of farmers' privilege to the fore. The story actually starts back in the mid-1980s, after the first plant was genetically transformed through GM techniques. Immediately following that breakthrough, a number of agro-chemical and seed companies began to investigate using this new technology to develop new proprietary seeds, initially tolerant to the new generation of proprietary chemicals. A watershed was reached in 1985. The US Patent and Trademark Office issued the world's first utility patent on a plant, the first private canola variety was registered in Canada and the glyphosate tolerant gene was isolated from a soil bacterium (and patented and trademarked as the Roundup Ready™ gene). As early as 1986 Monsanto began to search in fields in Western Canada for canola varieties that had a natural resistance to glyphosate, which Monsanto produced under the trade name Roundup™. By 1987 Monsanto had assembled the technology and a research consortium and had undertaken in collaboration with others the first transformations of canola to insert the Roundup Ready gene it owned. The first transformants were in field trials in Western Canada by 1988.

The 1990s saw the emergence of this new technology into the market place. Although utility patents on plants are not allowed in Canada, the decision in 1990 to adopt the Plant Breeders' Rights Act went part of the way to provide some sense of confidence in the industry that investments in plant varieties could be recouped in the Canadian market. On 23 February 1993 Monsanto Ltd. was granted Canadian Letters Patent No. 1,313,830 on a glyphosate-resistant gene for use in canola. That specific patent expires 23 February 2010.

By 1995 the Canadian government had approved Monsanto's RR canola variety RT73 (Canadian Decision Document DD95-02 by AAFC) and 5,000 acres were cultivated in Western Canada under a voluntary company-managed Identity Preserving Production and Marketing (IPPM) contract. IPPM production continued in 1996, with 50,000 acres of RT73 cultivated. (Meanwhile, AgrEvo introduced its own transgenic Liberty-Link™ variety, which went through similar regulatory and marketing steps). When Japan approved RT73 (and AgrEvo's competing varieties), the canola industry agreed to end the IPPM system for the new varieties and an estimated 450,000 acres of RR canola were planted in 1997.

Roundup Ready canola has been highly successful. In 1998, RR canola plantings rose to 2.6 million acres and in 1999 they peaked at 5.2 million acres. Since then Monsanto has averaged approximately 5 million acres of RR canola annually. The main reason the new variety was successful is that it conferred benefits producers. In the first instance, the new technology package (seed, chemical and royalty fee) was highly competitive with conventional seeding programs and the other new packages offered by AgrEvo and Pioneer Hi-Bred. Pioneer Grain Company Limited (cited by Phillips, 2003) calculated RR canola would cost on average $38/acre, compared with $43–47/acre for other programs. Recent research (Phillips, 2003) identified that producers generated an annual producer benefit of about $100 million in 2000: the new varieties reduced the amount and number of herbicide applications; delivered somewhat cleaner crops that faced lower dockage discounts; and allowed producers to plant their canola earlier, which improved yield by about 1.5 bu/acre (the benefits were partly offset by some yield drag, marginally lower canola prices and other management costs).

Monsanto used a wide range of IPR mechanisms to manage its new product. In the first instance, all of the processes used to develop the seed were patented or protected through trade secrets, the Roundup Ready gene itself was patented and the resulting variety was protected with a plant breeders' right and was branded with a trademark (Roundup Ready™). But because of the nature of canola—being open pollinated with high seed productivity (i.e. every seed can generate 200 progeny)— and Monsanto's limited proprietary position—the herbicide Roundup was off the patent—Monsanto decided to implement a growers contract. Before farmers were allowed to buy the seed, they were required to attend an information session and then sign a legal contract that specified they would only use seed bought from the trade, would buy and use only glyphosate produced and marketed by Monsanto (that provision was dropped in later years), follow certain reporting rules, keep detailed production records, sell all resulting harvest to the food trade and not keep any seed back for reuse. The contract also specified that Monsanto had the right to enter any contracting farmer's fields or operations to inspect for infractions of the contract, for three subsequent years.

Even though Monsanto insisted on farmers signing a TUA agreement before they could acquire seed, they remained concerned that some farmers may gain access through other means or renege on the terms of the contract. Monsanto established a "tip line" and reported it received in 1997 its first tip that Mr. Schmeiser was cultivating a RR canola crop without a license. Monsanto initiated a private investigation (documented in the court records) that continued through 1998. In 1999, Monsanto took action, acquiring samples of Mr. Schmeiser's crop and launching a civil lawsuit against Mr. Schmeiser and his farm operation, Schmeiser Enterprises Ltd. Mediation talks in 1999 between Mr. Schmeiser and Monsanto ended in failure and the case went to trial on 5 June 2000, with Monsanto arguing that Mr. Schmeiser grew Roundup Ready canola without a license. Mr. Schmeiser launched a countersuit against Monsanto (which he has yet to prosecute), arguing a variety of wrong-doings including libel, trespass, and contamination. The lower court ruled on 29 March 2001 that Mr. Schmeiser had used Monsanto's patented technology on 1,030 acres in 1998 without a license and levied a judgement against him for damages equal to $15/acre plus costs ([2001] F.C.J. No. 436). Mr. Schmeiser appealed that judgement to the Federal Appeal Court in 2001. On 4 September 4 2002, the three Justices of the Appeal Court unanimously upheld the verdict and findings of the trial judge ([2003] F.C. 165). Mr. Schmeiser then applied for and was given leave to appeal those verdicts to the Supreme Court of Canada.

Mr. Schmeiser and his legal team presented the Supreme Court with three key arguments related to farmers' privilege. First, they argued the PBR Act is the true reflection of Parliament's wishes, rather than the Patent Act. The gist of this point is that given that the PBR Act was passed and promulgated in 1990, after the emergence of plant patents in the US, Parliament was aware of the issue and chose to sustain farmers' privilege through PBRs rather than adopt a more restrictive IPR regime. Second, they argued that patents are not valid on living matter. They made reference to the Supreme Court of Canada decision on the Oncomouse and the US minority opinion on Chakrabarty. Furthermore, they argued that because Monsanto cannot fully control access to their open pollinated crop (there are numerous examples of cross-pollination in the field), their rights are extinguished through prescription. Third, they argued that even if patents exist, they conflict with the common law right to make use of whatever is naturally found on one's land—in this case the genetic material in seed that is reputed to have found its way naturally onto Mr. Schmeiser's field.

The Supreme Court brought down a mixed decision in May 2004 ([2004] 1 S.C.R 902). The Court first ruled that Monsanto's patent, on the RR gene construct and not the whole plant, was valid. Second, the Court confirmed its judgment from the Oncomouse case that patents are not allowed on multicellular organisms. Third, the judgment stated that in this case the fact that the RR canola in question expressed the proprietary RR gene cassette in all of the cells of the plant, that that conferred on Monsanto the right to exclude others from using any part of the plant. In essence, the court ruled that with the current state of technology (i.e. whole organism transformations) companies have the right to exclude others from accessing any part of those organisms that they have transformed with a patented gene construct. In essence, patents (even gene patents) trump the farmers' privilege and research exemption enshrined in PBR systems. Perhaps most puzzling to industry, in spite of the strong ruling in favor of the rights of patent holders, the Supreme Court overturned the monetary judgment of the lower court, arguing that the plaintiffs did not present any evidence that Mr. Schmeiser and his operations generated any profit on their canola crop, let alone benefited from the disputed technology. While this was most certainly a relief to the defendant, it does raise questions about what penalties and other disincentives patent holders can impose on those who use their technology without license.

While the legal issue appears for the time being to have been resolved in Canada, bringing the country more in line with other advanced industrial nations, it leaves unclear the practical effects of proprietary approaches and farmers' privilege.

THE PRACTICAL EFFECTS OF FARMERS' PRIVILEGE

There now is a clash of perspectives about how to support agriculture. On the one hand, inventors and biotechnology companies anticipate using the recently expended IPR systems to sustain new areas of research. On the other hand, many farmers and the international development community want greater freedom for farmers to access and benefit from new technologies.

Some have argued that legislation is needed to extend farmers' privilege into the patent systems. The Canadian Biotechnology Advisory Committee (CBAC, 2002), for example, recommended Parliament amend the Canadian Patent Act so that "a farmer's privilege provision be included in the Patent Act that specifies farmers are permitted to save and sow seeds from patented plants, or to reproduce patented animals, as long as these offspring are not sold as commercial propagating material, in the case of plants, or commercial breeding stock, in the case of animals." CBAC explicitly noted that while they support farmers' privilege, they respect the right of IP owners to enforce private contracts that could remove any privilege. While apparently logical, the fundamental flaw with this process is that allowing both farmers' privilege and private enforcement may simply raise costs of commercialization, which would act to lower investment without necessarily providing any effective privilege to farmers.

It is important to keep in mind that industry has a wide array of options if faced with incomplete or uncertain IPR mechanisms. First, industry might simply retrench and the public sector re-emerge as the main funder and developer of new germplasm and commercial varieties. While attractive to many, this is not likely to happen. The scale and complexity of using and commercializing products of biotechnology, a highly globalized science, effectively makes it difficult (if not impossible) for public breeders to develop new varieties (only two GM varieties have been developed and commercialized by public sector breeders—HT flax by the University of Saskatchewan [no longer in the market] and viral resistant papaya by Cornell University [commercialized but having difficulties being adapted and adopted to key markets]). Even if the will and financial support were available (which could conceivably require a tripling of public investment in agri-food research just to match current private research), the complexity of developing new varieties appropriate to market conditions could undercut the effectiveness of such an approach.

Second, some have suggested that industry may retrench and only develop varieties where effective biological controls are possible. A number of biological options exist, depending on the crop and its attributes (Smyth et al., 2002). Both traditional and molecular genetic methods already provide mechanisms to create hybrids (e.g. corn and canola) while working at a more refined molecular level offers the potential to more precisely control GM traits. Recently there has been an effort to reduce regeneration options by engineering foreign genes via the chloroplast instead of the nuclear genome. Such recombinants would only express the new traits in selected parts of the plant, rather than in the whole plant. Hence, any escape through pollen drift would not include the transgenes. Other options offer similar promise. In flowering plants, pollen is responsible for delivery of male gametes to the female carpel of the same or another plant, respectively resulting in self- or cross-pollination (Sawhney, 2001). It looks increasingly probable that pollination could be controlled through genetic manipulation of its development, structure or function—pollens with an incomplete set of genetic material would potentially impede pollination. The 1990s brought other efforts to ensure sterility, including Genetic Use Restriction Technologies (GURTS) that can turn off reproduction for either transgenic varieties or traits. Some dubbed this approach the "terminator gene." Patent data suggests that three competing systems are being worked upon. The use of sterile seeds has been widely practiced without much debate. The crudest form of this technology was known and is documented long before we fully understood genetics. Writings of ancient Egyptians and Greeks (c. 3000 BC) confirm that seedless grapes existed then. Since the 1930s, wide ranges of seedless edible crop varieties have been produced by traditional plant breeding methods (e.g. seedless grapes in 1936 and watermelons in 1951). Greater understanding of flower pollination, fertilization, fruit development, genetics and recombinant DNA-based technologies now enable scientists to apply this technology to new crops. Already we are seeing companies such as Monsanto and Bayer focus more of their energies on developing or commercializing new proprietary traits in hybrid systems or crops with biological barriers to reproduction.

Third, there is the potential that private research may abandon certain crop sectors, traits or markets. Even without the risks of uncertain property protection, there has been significant questioning in the industry about whether it can realistically expect to earn adequate returns in some of the smaller acreage crops. As recently as five years ago, most multinationals bragged that they were undertaking research on up to 10 different traits (e.g. herbicide tolerance, insect and viral resistance and abiotic stresses) in up to 20 different species. Given the increasingly uncertain regulatory environment and unpredictable consumer response, most firms now report that they have concentrated their energies and resources on at most four traits in five crops. Uncertain IPR protection could simply support this trend. Furthermore, firms have indicated both through action and word that they will treat markets differently depending on their level of IPR protections offered. For example, most companies have been very hesitant to commercialize their proprietary, open-pollinated transgenic crops in China and India, where property protections are either absent or only weakly enforced (this is clearly not the only reason for delayed introduction to those markets, as slow regulatory reviews have contributed to delays). Perhaps more telling is Monsanto's decision, announced 8 January 2004, that it would no longer sell RR soybean seeds in Argentina or carry out research to develop new varieties tailored to local Argentine conditions. Monsanto introduced its Roundup Ready soybeans there in 1996 and Argentine farmers enthusiastically adopted the crop, so much so that in recent years more than 95% of the soybean acreage in Argentina used RR soybean seeds. The difficulty has come in that Argentina has not offered any effective patent protection on the seed or its component parts, private contracts have not been possible and the local plant breeders' rights act has not prevented a blackmarket trade in proprietary seeds. The US Congressional Budget Office confirmed this in a recent study that showed that Monsanto was earning less than US$2/acre of RR soybeans planted in Argentina, compared with about US$8/acre earned over a comparable period in the US (US GAO, 2000). As companies gain experience in different species, traits and markets, they may become more selective on how and where they invest in new technologies.

Fourth, Monsanto has shown some innovation in attempting to use the provisions of TRIPS to negotiate country license and royalty schemes for their technology in markets where there are limited means to effect royalty schemes with individual producers. Monsanto recently has been negotiating with Argentina, Brazil and Uruguay to structure an end use license and royalty scheme for RR soybeans. These three markets have heavily adopted RR soybeans through informal and unlicensed transfers of seeds originally commercialized in Argentina. Current estimates are that almost all of Uruguay's soybean production uses RR seed and that 40% or more of Brazil's production uses similar seed. Effectively this system would involve royalty being calculated based on estimated acreage using the proprietary seed. A trade or some official agent at export position would then collect that amount through some form of export tax. The challenge of such a system, even if it can be made to work, is that it could be particularly inequitable if non-adopters have to pay part of the royalty.

Fifth, companies could simply choose to market their products differently. They could follow Monsanto's lead and only sell seed under private contract (i.e. current practice for RR crops), which means they would only offer their products in markets where private contracts are allowed and enforced. As noted in Table 1, some countries have limited provision for private contracts and many countries enforce contracts with varying vigor. The US has perhaps the most effective system for contracts as the courts there are willing to enforce the contracts by awarding both real and punitive damages in cases of contract infringement. In contrast, in Canada courts are hesitant to provide punitive damages, with the result that most breeches in contracts cost more to enforce than they would return. Other marketing options include long-term contracts, strategic alliances and a wide range of strategies to implement vertical integration either through merged production processes or equity investments. Alternatively, firms could simply change their pricing strategies. If they are unable through contracts, patents or other means to control farmers retaining and reusing seed, they may simply increase the price on the first season of introduction, so that they earn the full value of their invention in one year. A compatible corporate strategy would be to accelerate the rate of new varietal and trait introduction to make it more likely that farmers would buy new germplasm rather than reseed outdated varieties. The concern with each of these corporate strategies is that they would tend to shift the risks of new technologies (i.e. that they don't offer the benefits that they promise) away from the inventor and innovator and more towards farmers.

Ultimately, each of these options might contribute to efforts to preserve and sustain some or all of the current volume of research, development and commercialization of new technologies and crops. The difficulty is they each offer only a "second best" outcome. The best outcome would be clear and effective rules that offer the appropriate property protections to generate adequate (but not excessive) returns on investments. That balance is more likely to come from clear rules rather than from jockeying by governments and firms to cobble together make-shift property systems that are expensive to develop and enforce.

CONCLUSION

The Monsanto-Schmeiser case highlights a number of concerns with how we manage the trade-off between creating incentives for private investment in R&D and ensuring optimal social benefit from any resulting inventions and innovations. As with most real-world issues, there is no unambiguous solution that leaves everyone completely satisfied. In absence of the public sector re-engaging as the lead funder, developer and proprietor of agri-food innovation, we are going to have to accept either some degree of private monopoly or accept lower private investment (which might impede growth in productivity). While some may prefer to see slower innovation, the social implications can be substantial, both for consumers living at or below the subsistence level and for producers who have to compete with other sectors for land, labour and capital. Ultimately, ignoring the problem will simply compound the costs. There are a number of potential conclusions to draw from this analysis of the issue of farmers' privilege.

First, it is worth remembering that the debate over IPRs and farmers' privilege is partly the result of different assumptions about the role of private property itself. There is a profound disagreement over whether patents are simply a tool of one power elite or whether they are an instrument for social good. To a great extent, this difference of opinion is the product of different value systems, which means that it will not be resolved by reasoned argument. We simply will have to agree to disagree on the starting point. If we can do that, then there may be room to debate how we can improve the operation of IPR systems to mitigate the costs and optimize the benefits.

Second, those engaged in the debate about farmers' privilege must keep in mind that patents and other legal instruments are only one of a wide range of potential IPR mechanisms that can enforce exclusion. There is a real risk that if we ignore the other systems—for example firm strategy, contracts and biological control systems—that we may design what appears to be the optimal patent or PBR system, but that it will not achieve its goals due to unanticipated interactions with the other mechanisms. Essentially, it is unwise to think in narrow, comparative static terms—the IPR system is wide ranging and dynamic (Phillips, forthcoming).

Third, it would appear that farmers' opportunity to sow crops with saved seeds has been and is currently more of a privilege than a right. There is only a very recent history of explicitly permitting farmers to save and reuse seed that was developed by a commercial enterprise. And this privilege is inextricably tied to an equity argument, which may be harder to sustain for many farmers in developed nations. Essentially, the FAO process has justified the formalizing of privilege on the basis that farmers individually and collectively (and without formal acknowledgement or compensation) are integral parts of the indigenous research process and vital custodians of biodiversity. While this may be plausible in developing nations where most farmers use traditional landraces and there is a wide diversity in their operations, this is likely less true in North America, Europe and Australia, where most farmers either use store-bought seed that comes from a commercial seed company or use seed saved from such a transaction. Furthermore, those farmers who are part of the seed industry in North America (e.g. doing trials or multiplying seed) are usually formally acknowledged and compensated for their efforts. Hence, the privilege is perhaps less justified in developed country markets than in LDCs.

REFERENCES

Canadian Biotechnology Advisory Committee (CBAC) (2002) Patenting of higher life forms: a report to the government of canada biotechnology ministerial coordinating committee (Final report)

Canadian Intellectual Property Office (CIPO) (2004) Patents. Available online: http://cipo.gov.ca

Centre for intellectual property (CIPP) (2004) Quarterly Newsletter, October. Available at: http://www.cipp.ncgill.ca/en/news/newsletter

Food and Agricultural Organization (2004) Commission on genetic resources for food and agriculture. Available online: http://www.fao.org

Nottenburg C, Parday PG, Wright BD (2003) South-North trade, intellectual property jurisdiction, and freedom to operate in agricultural research on staple crops, *Economic Development and cultural change* *51*:309-336

Phillips P (2003) The economic impact of herbicide tolerant Canola in Canada. In Kalaitzanondakes N (ed) Economic and environmental impacts of first generation Biotechnologies. CABI International, Wallingford, Oxon, UK.

Phillips P (Forthcoming) Technology, ownership and governance: an alternative view of IPRs. In: Einsidel E (ed) Foresight on emerging technologies. University of Calgary Press, Calgary.

Phillips P, Khachatourians G (2001) The Biotechnology revolution in global agriculture: innovation, invention and investment in the canola sector. CABI Wallingford, Oxon, UK

Sawhney VK (2001) Pollen Biotechnology. In: Khachatourians GG, McHughen A, Nip W-K, Scorza R, Hui Y-H (eds) Transgenic plants and crops. Marcel Dekker Inc New York

Smyth S, Khachatourians G, Phillips P (2002) The case for institutional and biological mechanisms to control GM gene flow. Nat Biotechnol 20(6):537–541

Traxler G (2003) Let them Eat GM? The Biotechnology picture in Latin America. Presentation to the conference Crossing over: genomics in the public arena, Kananaskis, Alberta, April 25

US Patent Office (1999) Search of patent database undertaken on http://patent.womplex.ibm.com, October 9

The International Union for the Protection of New Varieties of Plants (UPOV) (1991) The act of international convention for the protection of new varieties of plants of December 2, 1961, as Revised at Geneva on November 10, 1972, on October 23, 1978, and on March 19, 1991. Available online: http://www.upov.int/eng/content.htm

The International Union for the Protection of New Varieties of Plants (UPOV) (1999) About UPOV. Available online: http://www.upov.int/eng/content.htm

US General Accounting Office (2000) Biotechnology: information on prices of genetically modified seeds in the United States and Argentina. Available online: www.gao.gov/archive/2000/r400055.pdf

US Patent and Trademark office. 2004. Definition of a utility patent. Accessible on line at: http://www.uspto.gov/main/glonary/#u

World Intellectual Property Office (WIPO) (2004) About WIPO. Available online: http://www.wipo.int

World Trade Organization (WTO) (2004) Intellectual Property (TRIPS): FAQs. Available oline: http://www.wto.org

CASES

Diamond v. *Chakrabarty* (1980) 447 US 303

Harvard College v. *Canada (Commissioner of Patents)* (2002) 4 SCR 45; 2002 SCC 76, (also known as the Harvard Oncomouse case)

Monsanto Canada Inc. v. *Schmeiser* (2004) 1 S.C.R. 902

Monsanto Canada Inc. v. *Schmeiser* (2003) F.C. 165 (CA)

Monsanto Canada Inc. v. *Schmeiser* (2001) F.C.J. 436

DAVID CASTLE AND E. RICHARD GOLD

CHAPTER 4

TRADITIONAL KNOWLEDGE AND BENEFIT SHARING: FROM COMPENSATION TO TRANSACTION

INTRODUCTION: GLOBAL WELFARE INEQUITIES

Eradication of the persistent inequities in the health, food security and environmental quality faced by people around the world remains one of the greatest challenges facing humanity. Scientific discovery and technological innovation play a fundamental role in the eradication of these inequities. In particular, genomics and biotechnology may contribute the greatest advances in reaching health, food and environmental development goals by way of ground-breaking technologies with global uptake, and by finding context-sensitive solutions to local problems. Confident predictions about biotechnology's capacity to change lives for the better is, however, mere soothsaying unless they are backed by realistic conceptions of technological development, actual products and services, coupled with a consideration of the social factors that condition the acceptance of new technology. These social factors help in determining the new technology's ability to redress welfare inequities.

Here we concentrate on one of the increasingly prominent aspects of the discourse over traditional knowledge and welfare inequities: the contention that the current failure to share financial and other benefits arising from biotechnologies derived from traditional knowledge is tantamount to theft and should be compensated. Two aspects to this contention have shaped current thought about the sharing of benefits. The first aspect is an allegiance to the idea that traditional knowledge is a unique form of knowledge that, in part because of its purported uniqueness, can be owned and protected (a view not so distant from common defenses of intellectual property protection in industrialized countries). The second aspect is the idea that exploitation of traditional knowledge by non-owners is inherently unjust and deserves remedy. From this normative underpinning derives the interpretation of benefit sharing as a form of compensatory justice.

In this chapter we challenge the conception of traditional knowledge and the compensatory approach to justice that usually undergird benefit sharing. The first part of our argument points out that apart from traditional knowledge being knowledge of special provenance, it is no different in kind than any other knowledge. This observation is relevant to understanding the problematic nature of property claims in traditional knowledge. We argue that while traditional knowledge may be historically localized within an ethnically or geo-politically defined group of people, this provenance does not lead to the conclusion that it ought to be protected through property rights. Consequently, if there is no property right in traditional knowledge, then nothing can be wrongfully taken by a third party and a claim based on compensatory justice cannot succeed—no compensation is due. Of course, one cannot overlook the fact that proprietary rights in traditional knowledge and compensation for its use can be considered separately—one could still endorse a compensatory approach to benefit sharing after having raised doubts that traditional knowledge is proprietary in virtue of being traditional. In anticipation of this tack in the argument, we conclude that whatever one's stance is with respect to traditional knowledge, benefit sharing should not be modeled on compensatory justice.

We instead argue that distributive justice provides a sounder basis for conceiving of sharing the benefits arising from biotechnological innovation. Distributive justice would have us allocate benefits arising from biotechnological innovation on the basis of our shared humanity. Distributive justice provides key advantages over a compensatory scheme. First, it does not depend on proprietary claims. Rather, it relies on fundamental principles underlying liberal democracies about fairness and access to resources. Second, because it does not rely on property claims, it does not require proof of past wrongdoing, thus liberating claims to share benefits from long discussions of past conduct. Third, as it is not historically contingent, it calls for the sharing of benefits with all of the worlds' poor countries and not simply those rich in traditional knowledge.

We undertake our analysis and discussion in the following order. First we define traditional knowledge and how it is wrongfully exploited that in turn gives rise to the call for benefit sharing. We then consider benefit sharing as compensatory justice and discuss the problems that arise with this approach. Finally, we present an alternative model of benefit sharing that is consistent with a formal theory of distributive justice. This position, we argue, indicates that benefit sharing agreements must be negotiated on a multi-lateral basis.

TRADITIONAL KNOWLEDGE AND ITS WRONGFUL EXPLOITATION

The World Intellectual Property Organization (WIPO) defines traditional knowledge very broadly as follows:

> ...tradition-based literary, artistic or scientific works; performances; inventions; scientific discoveries; designs; marks, names and symbols; undisclosed information; and all other tradition-based innovations and creations resulting from intellectual activity in the industrial, scientific, literary or artistic fields... (WIPO, Intergovernmental Committee on Intellectual Property and Genetic Resources, Traditional Knowledge and Folklore, 2002).

For the purposes of sharing benefits from scientific and biotechnological innovation, WIPO has confined this definition to "knowledge about products or processes, natural or artificial, that are relevant to biotechnology innovation, and known by some people but not all." This definition is consistent with the Convention on

Biological Diversity (CBD) that explicitly relates traditional knowledge to benefit sharing:

> Article 1: The objectives of this Convention, to be pursued in accordance with its relevant provisions, are the conservation of biological diversity, the sustainable use of its components and the fair and equitable sharing of the benefits arising out of the utilization of genetic resources, including by appropriate access to genetic resources and by appropriate transfer of relevant technologies, taking into account all rights over those resources and to technologies, and by appropriate funding.
>
> Art. 8(j): Subject to its national legislation, respect, preserve and maintain knowledge, innovations and practices of indigenous and local communities embodying traditional lifestyles relevant for the conservation and sustainable use of biological diversity and promote their wider application with the approval and involvement of the holders of such knowledge, innovations and practices and encourage the equitable sharing of the benefits arising from the utilization of such knowledge, innovations and practices.

From these definitions, one can notice two types of traditional knowledge that relate to biodiversity and innovation. First, there is knowledge about the use of biological resources in health or agriculture based on previous exploitation. Well known examples include the medicinal uses of the neem tree and of turmeric. Second, there is the information about the existence of particular plants or animals that have characteristics that may be of interest to a pharmaceutical company. A good example of this is Costa Rica's establishment of the Instituto Nacional de Biodiversidad (INBio) that enters into partnerships with companies wishing to explore the country's biodiversity (see chapter 9 for more discussion).

Calls for sharing of the benefit derived from the exploitation of the two above mentioned types of traditional knowledge are often modelled upon the following wrongful exploitation scenario. An indigenous group has traditional knowledge. Another group, typically but not necessarily members of an industrialized country, recognizes the potential utility of the knowledge and exploits it. When the latter does so, it gains access to and control over the benefits arising from the knowledge to the exclusion of the indigenous group. As a result, an objection is raised that this is an inequitable outcome. At the very least, it is deemed preferable if all could benefit in the exploitation of knowledge and have equal access to it. The situation is made more egregious when the industrial party asks for compensation from developing country consumers for goods and services incorporating the exploited knowledge. This possibility resembles extortion.

At least since the negotiation of the CBD in 1992, the international community has recognized wrongful exploitation and struggled to conceive of how to allocate both access to and the benefits derived from traditional knowledge. Article 8(j) of the convention aspires to protect this knowledge in some form without giving substantive content to the form of protection required. Subsequent work has led to the development of Bonn Guidelines on Access to Genetic Resources and Fair and Equitable Sharing of the Benefits Arising Out of their Utilization (Secretariat of the Convention on Biological Diversity, 2004) setting out voluntary guidelines on access to genetic resources which state that "[m]utually agreed terms could cover the conditions, obligations, procedures, types, timing, distribution and mechanisms of benefits to be shared."

The meaning of "traditional" often simply locates knowledge in a set of historical and geographical contingencies—where the knowledge was used and by whom. Yet, however banal these contingencies might seem from an epistemological standpoint, they constitute the strongest evidence for wrongful exploitation of traditional knowledge. The agglomeration of modern seed banks in industrialized countries drawing on resources from the developing world, for example, can be seen as a modern outgrowth of traditional colonial practices of bio-geographical exploitation (Juma, 1989). Place and practice are likewise implicit in terms like "ethnobiology" and "bioprospecting" currently used to describe the study of people's local knowledge in the first case, and utility interests in that knowledge in the second case.

Benefit sharing as compensatory justice is premised on the belief that traditional knowledge gives rise to special property rights. There are two primary arguments put forward that justify treating traditional knowledge as property. First, there is the frequently adopted position that traditional knowledge is knowable by certain distinct groups but not necessarily by all people and so, traditional knowledge demarcates a way of knowing. This unique way of knowing gives rise to a property claim that ought to be compensated for if wrongfully exploited. For example, Vandana Shiva states that people in developing countries (India in her case) understand nature differently than do people in the West (Shiva, 1997). She finds evidence for the different epistemological outlooks in the vulnerability of Eastern cultures to biopiracy. Because people in these cultures are not inclined to think of nature as being a resource for technological extraction and the object of intellectual property protection, they are exploited (Shiva, 2000). Shiva draws a strong contrast between Eastern and Western ways of knowing to leverage injunctions against the use of developing country genetic resources in Western biotechnological innovation. Further, she claims that the unique ways of knowing in the East give rise to a proprietary claim over traditional knowledge. By treating traditional knowledge as a form of property, we can curb the incursion of Western biotechnology in Eastern contexts. In this way, differences in epistemological stance lead to the protection of Eastern proprietary knowledge.

The second approach put forth to justify treating traditional knowledge as property is to identify the special nature of "traditional" knowledge that imparts a special status on its knower. The special status of "traditional" knowledge focuses on whether the idea of "traditional" imparts some particular status on the knowledge or knower and so gives rise to a property right. This approach is aimed at the obvious tension between the push to protect proprietary interests in traditional knowledge and underlying cultures which have not yet come to hold this knowledge to be "ownable" or exploitable.

Standpoint epistemology, of the kind advanced by Sandra Harding, has lent credibility to the idea that there are different ways of knowing attributable to groups of people with unique socio-economic circumstances (Harding, 1986). How one relates to one's world may be subject to interpretations nuanced by place and culture. One can go further and claim that localized knowledge constitutes not a different set of knowledge but a different way of knowing. This seems to be a good starting point for explaining why "traditional" knowledge is special. Unfortunately, it does not get us very far in the case of traditional knowledge.

Standpoint epistemology may be well and good when the question is whether one group knows different than another, but that is not the centre of concern relating to benefit sharing. By definition, the call to share benefits relates to knowledge that *is* known outside the community of origin. Otherwise, that knowledge could not be exploited. If knowledge was truly not understandable by others outside the community, there would be no need for property rights since the community, by its unique ability to understand, effectively excludes the rest of us from using that knowledge. It is only where knowledge can be understood and appreciated by others that we need a property regime to (artificially) determine who ought to be able to exclude others.

Thus, a claim that there are different ways of understanding one's natural surroundings is not germane to deciding whether "traditional" knowledge ought to be protected through property rights. The claim can instead be seen as an attempt to deflect the issue away from a discussion of the kind of property rights that might be granted in respect of traditional knowledge to the idea that traditional knowledge is something unique and provides inherent rights to exclude that knowledge in Western science. In effect, the move to reify traditional knowledge is an attempt to generate protection for some knowledge by claiming that it is part of a different and possibly radically non-translatable conceptual scheme (Davidson, 1984). This is an implausible although often nakedly asserted claim about knowledge when in fact the real issue is who had the knowledge and in what priority. After all, what makes traditional knowledge valuable and worth conserving is that it is knowledge already found to be useful (e.g. the medicinal properties of the neem tree or of turmeric), which means that it has already been exploited. So questions about *whether* it should be exploited and if it *can* be exploited are, ultimately, beside the point. Instead, the question is *who* should have access to the knowledge and *who* should be entitled to exploit it. To conclude this discussion, it follows that the move to reify traditional knowledge introduces epistemological tensions that do not resolve the question of ownership. Rather, this approach merely obfuscates the grounds upon which knowledge localized to people and places may be said to be traditional.

BENEFIT SHARING

Benefit Sharing and Compensatory Justice

Benefit sharing is a policy-level decision about how the acquisition and use of traditional knowledge related to biotechnology should be handled. In this respect, benefit sharing has no specific normative underpinning of its own but is appropriately regarded as simply a mechanism through which to implement some externally-determined obligation. As we noted above, benefit sharing is claimed to derive its normative basis from the value of traditional knowledge itself. It then establishes a framework that seeks to allocate this value between those from whom the knowledge derives and those who exploit it. This framework provides such international organizations as the Food and Agriculture Organization (FAO) and the WIPO with benchmarks for the distribution of financial and other benefits to individuals or to populations based on their provision of material or samples.

Given that benefit sharing merely establishes a framework, the obligation to share benefits is constructed differently depending on how we conceive of traditional knowledge. If traditional knowledge is subject to proprietary claims, then benefit sharing amounts to a claim for compensation for the wrongful or unauthorized use of that knowledge. If on the other hand, we do not conceive of traditional knowledge as proprietary, then benefit sharing cannot find a normative basis in any special nature of traditional knowledge; it will have to find it elsewhere.

Nevertheless, international agreements appear to accept the link between benefit sharing and proprietary claims to traditional knowledge. One can trace its foundations back to at least the 1992 CBD that, once in effect in 1994, acknowledged each State's sovereign right over its genetic resources. Benefit sharing was a major agenda item at the fourth Conference of the Parties that called for the articulation of an international framework to satisfy Article 15 of the CBD that provides for international access to genetic resources under national control (UNEP/CBD/COP/4/21-3). Following on the CBD, the Bonn Guidelines (adopted in 2001) established procedures to protect the rights of traditional knowledge holders while enabling access to this knowledge (UNEP/CBD/COP/6/20).

Access to resources has become conditional on obtaining prior informed consent and, increasingly, on the entering into of contractual obligations, particularly as the Parties move toward harmonization of the CBD with TRIPS and WIPO agreements. Traditional knowledge also has a place in human rights discourse, specifically under the UN Commission on Human Rights (UN Commission on Human Rights, High Commissioner, 2001). Yet, despite the recent nuances given to the term, benefit sharing continues to follow the compensatory model. At the seventh Conference of the Parties held in Kuala Lampur in 2004, the Executive Secretary to the CBD continued to reflect the compensatory approach in his opening remarks when he called for "the equitable sharing of benefits arising from the utilization of genetic resources, which was linked to other issues such as the recognition of and fair compensation for the utilization of the traditional knowledge." (Ad hoc open-ended working group on access and benefit-sharing on the work of its second meeting, 2003).

Expanding beyond traditional knowledge, we see a similar compensatory link between benefit sharing and human genetic information. Perhaps the clearest articulation of the compensatory approach arises in the Human Genome Organization's (HUGO) *Statement on benefit sharing*. In that statement, HUGO endorsed a model of benefit sharing substantively based on compensatory justice that it defined as "meaning that the individual, group, or community, should receive recompense in return for [a] contribution..." (Human Genome Organization HUGO, 2000; Sixth Meeting of the Conference of the Parties to the Convention on Biological Diversity, 2002). While HUGO's *Statement* refers most clearly to human genetic information, its compensatory approach to benefit sharing has become paradigmatic for not only human genetic information but also traditional knowledge about biodiversity. It is also worth noting that there is a strong current of compensatory justice thinking running throughout the literature on ethnobotany, biopiracy and benefits sharing (Downes, 1997; Downes, 1999; Downes, 2000; Nijar, 1996).

Critique of the Compensatory Approach

If international agreements, guidelines and the literature accept that benefit sharing is based on a compensatory claim to the unauthorized taking of proprietary traditional knowledge, it should be easy to articulate the reasons for which this knowledge is proprietary. It turns out, however, that the reasons are more illusory than real. Based on the foregoing, one may be tempted to start with the point that "traditional" identifies knowledge according to place, history or culture to argue that this specific origin gives rise to a proprietary interest by those sharing place, history and culture. As we argued earlier, such an argument cannot be sustained. That is, one cannot coherently claim that traditional knowledge that has already been exploited and that can be understood (for purposes of further exploitation) by others is epistemologically so different than other forms of knowledge so as to justify creating a property right. Thus if traditional knowledge is to be subject to property rights, the justification for this must be found elsewhere than in its special nature.

Our starting assumption is that, in a liberal society, goods ought to be free for anyone to use unless a convincing argument can be given to limit that freedom (Macpherson, 1978; Hettinger, 1989). This assumption is open for dispute, although we notice that it informs the aspirational dimension of international agreements on benefit sharing. This is particularly true of intangible goods such as knowledge that is non-rivalrous and even more so of global public goods. Property rights detract from this presumptive state of freedom, and therefore must be justified on some normative basis. Thus if traditional knowledge does not give directly rise to a proprietary claim because of its distinct nature, those in favor of claiming a property interest in such knowledge have the onus of convincing the rest of us of their claim. The usual basis for such an argument is either a labor theory of property or utilitarianism.

Those arguing in favor of proprietary claims to traditional knowledge often do so based on the Lockean argument that where one has labored, one ought to receive a property right. Locating this natural property right in traditional knowledge follows the Lockean model: Those who have property in themselves and mix labor with the natural resources and thus have dominion over the property, thereby have the right to prevent appropriation of their property by others. Once we accept that traditional knowledge gives rise to a natural property right, benefit sharing gains a very specific normative underpinning associated with property rights. In the paradigmatic scenario described earlier, appropriation and exploitation of traditional knowledge by third parties violates the property holder's right to exclude all others from that knowledge and so constitutes theft. Although "biopiracy" may sound metaphorical, in these cases, it is meant literally.

Unfortunately for those who put forth the argument, not only is there controversy about the general Lockean argument, but its application to traditional knowledge is deeply problematic for two reasons. First, traditional knowledge has, by its very nature, been around for a long time and so this means that nobody using it today labored to create it (if they had, they would already have a claim to traditional knowledge on standard legal grounds). Thus, these individuals do not have a direct claim based on the expenditure of their labor. Second, relying on the labor of one's ancestors and a subsequent assignment of the property right to the present generation is similarly unpersuasive. In fact, it is an argument that has never held sway, even in respect of inventions and artistic works. We grant no absolute rights in ideas or artistic works beyond a limited time period. As all creative works receive fairly limited rights, those rights would have long expired in respect of the very large bulk of traditional knowledge. Consequently, no persuasive Lockean argument to property in traditional knowledge exists.

If we cannot justify a property right in traditional knowledge on the basis of labor, another justification to consider is utilitarianism. According to utilitarianism, the granting of a property right is justified so long as it increases benefit to society. The property system covering intangibles most associated with utilitarianism is the patent system. The central justification for the award of patent rights is the utilitarian argument that patents lead to the creation and dissemination of knowledge (Hughes, 1988). In particular, patents provide a financial lure to would-be inventors to invent new things. Without this lure, inventors may not undertake the risk and expense involved in developing such knowledge. As Gold and Caulfield have pointed out, no such arguments exist with respect to traditional knowledge (Gold and Caulfield, 2003). By its very nature, traditional knowledge has been around for a long time, having been built up over many years. Granting a property right will provide no incentive to invent as the invention has already taken place. Patent systems have long recognized this. In fact, the English Crown's failure to accept it led to the passage of the Statute of Monopolies (Walterscheid, 1994). Patent systems thus apply only to "new" knowledge, excluding what was previously known (Leskien and Flitner, 1997).

The conclusion is therefore that neither natural property rights nor utilitarian arguments justify a property right in traditional knowledge *per se*. This does not mean that no traditional knowledge may become subject to intellectual property rights on the same basis as other knowledge; it simply means that the mere fact that knowledge is traditional is insufficient in itself to justify the granting of a proprietary interest. In the absence of a justification to limit freedom, we can only conclude that we are free to use traditional knowledge as we wish.

Furthermore, the attempt to justify proprietary rights in traditional knowledge only perpetuates an ideology of proprietarianism that restricts access to other forms of knowledge, such as that embodied in medications and agricultural research. This is because the proprietary claim to traditional knowledge mirrors the strongest form of argument in favor of intellectual property rights including patents. Increasingly, international discourse over intellectual property has taken on the tones of this property being the natural right of its holder with any limitation on this right being *prima facie* illegitimate (Drahos, 1996). By reiterating this argument, even in relation to traditional knowledge, we not only condone but lend support to the creeping proprietarianism at the international level. As whatever practical benefits developing countries may derive from treating traditional knowledge as property will be far outweighed by a further entrenchment of proprietarianism with respect to patents, the move to argue that traditional knowledge is property can only be viewed as dangerous.

As Goes the Property Argument, So Goes Benefit Sharing

As discussed above, the claim to a property right over traditional knowledge currently grounds compensatory justice based benefit sharing schemes. The problem is that since the property right claim is incoherent in the case of traditional knowledge, the grounds for claiming compensation are negated. In this scenario, bioprospecting cannot lead to biopiracy. On its own internal logic, a benefit sharing mechanism should provide no remedy to a third party's use of traditional knowledge since there can be no harm done to the alleged holders of an inexistent property right. Nevertheless, it would be wrong to simply ignore the strength of the discourse linking *"traditional"* knowledge as property with benefit sharing. Acknowledging the strength of this discourse is, however, different from accepting it wholeheartedly. In fact, if we look further into the traditional knowledge debate, we can identify different forms of argument. One of these is the proprietary argument addressed above. A second, however, is fundamentally about credit, not property.

A useful starting point is to recognize that the value of traditional knowledge—and the reason it is the source of dispute—is because it can be exploited. Were this not true, there would be no need for discussions about benefit sharing. The existence of bioprospecting in itself provides a tacit acknowledgement that traditional knowledge is valuable in relation to the genetic resources it makes available. Biopiracy occurs when the bioprospector acts as if the traditional knowledge that was valuable in identifying the genetic resource either does not exist or was not exploited prior to the actions of the prospector. This difference between bioprospecting and biopiracy manifests itself, for example, in disputes about whether countries in receipt of traditional knowledge recognize "foreign prior art." In the United States, for example, recognition of traditional knowledge as prior art is variable. Where the USPTO has recognized that there was foreign prior art in the form of traditional knowledge, it has revoked patents (e.g. revocation of the 1995 patent on the use of tumeric in wound healing and the 1986 revocation of the patent on ayahuasca). On the other hand, US law does not explicitly require that foreign prior art be considered—except in cases where a patent claim has been filed or use of the invention explicitly documented—which leaves unclear the extent to which the sections 102 and 302 that refer to the conditions for patentability are intended to cover foreign and not just domestic prior art. The practical consequence is that US patents can be filed permitting the exploitation of traditional knowledge, regardless of whether the patent will ultimately withstand a novelty challenge. This means that prospectors can claim credit for knowledge which they have not themselves created. It must be pointed out, however, that the remedy to this problem of credit-taking is not the grant of property rights over traditional knowledge, but the *absence* of property rights over the claimed invention by the prospector. That is, the argument here is not one related to benefit sharing but to ensuring high levels of public access to genetic resources.

Overall, we can therefore conclude that if the hope was to provide compensation for the use of traditional knowledge by constructing a benefit sharing scheme based on a property platform, the reality is that not only will no compensation flow to those possessing such knowledge, but access to other knowledge, medicines and goods may become even more difficult to gain. Oddly, then, we find ourselves in a position where the ostensibly laudable goal of benefit sharing is undermined by the very strategy to secure it. Is there an alternative?

BENEFIT SHARING AS DISTRIBUTIONAL JUSTICE

If compensation is not a viable principle on which to base benefit sharing, a different principle of justice must undergird the concept. Recognizing that welfare inequities constitute the greatest challenges facing humanity, the starting point for creating a benefit sharing framework should be the need to seek greater distributional equity in a proactive fashion. While past harms must be redressed, distributional equity must extend beyond retrospective assessments of how traditional knowledge was exploited.

Consequently, instead of basing obligations to share benefits on the basis of something *taken*, we can construct it as a mechanism to distribute goods that are of interest to all. That is, instead of conceiving benefit sharing within a compensatory justice framework, we can see it as a means through which to achieve distributive justice with respect to such global public goods as health care products and services. On this view, we give form to benefit sharing through an examination not of traditional knowledge but of rights to health and to scientific progress. In fact, traditional knowledge becomes irrelevant to the obligation except, perhaps, as an argument that populations with traditional knowledge have a positive obligation to share it, at least when appropriate regimes exist to share the benefits of scientific research.

The argument underlying the obligation to share benefits can be constructed on both theory and positive law. Rawlsian notions of justice would lead to a distribution of such foundational goods as health and food (Rawls, 1999a). Alternatively, one could ground the obligation to share benefits on international human rights instruments, such as the *Universal Declaration of Human Rights* and the *International Covenant on Economic, Social and Cultural Rights*. Both provide the foundations for not only a right to health, but the obligation of all states to ensure that the most disadvantaged people have access to health care (Chapman, 2002). In either case, benefit sharing would simply be a mechanism through which to achieve either distributive justice or to satisfy international human rights law.

Rawls (1999b), Beitz (1978) and Pogge (1994) all present views that extend theories of distributive justice to the international realm. However, these global distributive justice principles have not been greatly elaborated. Benefit sharing thus remains a vague concept except in one respect: it is an obligation owed to all peoples regardless of their ability to provide traditional knowledge (de Ortuzar, 2003). The obligation to share benefits derives from one's status as a person and not from one's control over certain knowledge. Consequently, we can completely break the special link currently made between *"traditional"* knowledge and benefit sharing. Benefit sharing, in these circumstances, becomes a mechanism to ensure the equitable distribution of both scientific research capacity and gains arising from scientific research, at least in the health care and agricultural fields.

Multi-Lateral Approach to Constructing a Distributive Justice Framework?

Benefit sharing is a policy tool designed to achieve certain normative outcomes from the interaction of nations and peoples and the exchange of knowledge and material. In itself, it neither specifies the goal of benefit sharing nor the principle of distributive justice upon which it rests. The absence of a clear normative base may explain why some have adopted the compensatory approach to benefit sharing. This approach speaks most directly to grievances about biopiracy and colonialist misappropriation of biodiversity. Under this paradigm of benefit sharing, disputes are bi-lateral and require bi-lateral remedies. While the specific normative objectives provided by benefits sharing schemes are properly the outcomes of international negotiations, the outline of a framework to define distributive justice in benefit sharing can be provided in advance. This framework is multi-lateral.

The High Commissioner of the UN Commission on Human Rights found three clear links between the TRIPs agreement and human rights that could provide the outline of a framework to define distributional justice. First, nations may uphold obligations under Article 15 of the International Covenant on Economic, Social and Cultural Rights (ICESCR) and the Universal Declaration of Human Rights by putting *ordre public* limitations into their patent laws (TRIPs Article 27(2)). Second, compulsory licensing provisos within TRIPs enables member states to balance public and private interests provided under Art. 15. Third, the High Commissioner regards TRIPs as enabling and encouraging cooperation between states, particularly industrialized country technological capacity strengthening in developing countries. Nevertheless, however much human rights may be formally accommodated by TRIPs, the Commissioner observed that "while the Agreement identifies the need to balance rights with obligations, it gives no guidance on how to achieve this balance" (para 4). In addition, TRIPs contains no specific provisions for protecting traditional knowledge that, when given the dominance of the US patent system and the problems noted above, raised criticisms that traditional knowledge is vulnerable to external control. These formal shortcomings of TRIPs suggested to the High Commissioner that the administration of TRIPs requires a normative underpinning. To some, this is a deficiency in the very nature of TRIPs; to others, it is an inevitable outcome of a negotiated agreement, and one that creates an opportunity to implement policy instruments like benefit sharing.

According to our model, global welfare equity is the over-riding objective and so benefit sharing must be anchored in a system that fosters distributional equity. Again, the precise nature of a good distribution is a matter for negotiation, but the adoption of a principle of distributional equity suggests how sharing of benefits arising from traditional knowledge should be handled. Noting that some developing countries may themselves unfairly exploit knowledge generated by indigenous groups (Sarma, 1999), some have argued that we can only achieve a fair distribution of benefits if we use international law and institutions to govern the exploitation of traditional knowledge. Similarly, there have been calls for a global bio-collecting society that would standardize collections, collecting procedures, and harmonize these with domestic intellectual property regimes, the CBD and TRIPs (Drahos, 2001). The central idea being proposed in all of these initiatives is that negotiating the exploitation of traditional knowledge within international agreements provides the most secure basis for benefit sharing. Unlike the compensatory model, where justice involves redressing past harm, a distributive approach involves negotiating the nature of a just distribution and the means of achieving that distribution between rich and poor countries.

Bilateral negotiations about benefit sharing can be productive. Since 1989, for example, INBio in Costa Rica has established the terms under which bioprospecting is conducted (www.inbio.ac.cr). More recently in Peru, the Aguaruna negotiated with Searle, an American pharmaceutical company, access to their traditional knowledge (Tobin, 1997). Some corporations are embarking on ambitious benefit sharing and access schemes that return training, infrastructure, IPRs and monetary benefits to countries of origins but only after contractual agreements have been established (www.diversa.com). These examples are early proofs that distributional equity achieved through negotiated agreements provides more equitable benefit sharing outcomes than is possible through simple retroactive compensation. However, it is unclear whether such bilateral agreements actually achieve distributional justice in a fundamental way.

Expecting rich and poor countries to negotiate just distributions of food and health is nice in theory, but *realpolitik* of power looms large. While the CBD sets up a mechanism of bi-lateral arrangements between resource-rich (but usually poor) countries and those wishing to exploit those resources, it fails to take into account power differences. The world of international trade has already provided us with significant evidence that bi-lateral negotiations between countries, particularly between the US and countries wishing to accede to the WTO, have been anything but fair (Trosow, 2003). In these circumstances, expecting rich countries to provide substantial benefit to poorer countries may be too idealistic.

A multi-lateral organization such as the WTO may be more successful in negotiating a distributive justice framework. As Adams suggests, poorer countries can at least compensate for *realpolitik* by using existing international institutions, most notably the WTO (Adams, 2002). While counterintuitive, given the common perception that the WTO agreements are unfair to developing countries, her argument is that only through the WTO can developing countries secure real benefits from developing countries since they can both hold up negotiations (and the benefits they provide to developing nations) unless their needs are met. The point here is simply that international negotiation is needed to define both what distributive justice requires and to craft benefit sharing mechanisms that achieve this distribution. While bi-lateral agreements have done some good, we believe that a multi-lateral approach, most likely under the auspices of the WTO, provides the most promising route to achieving justice.

CONCLUSION

Justifications for benefit sharing cannot be derived from claims to property rights in traditional knowledge, if not because natural property rights are themselves problematic, then because property is normally considered free unless there is a normative justification for restricting access, particularly in the case of knowledge assets. But the focus of this chapter is less on these considerations and more on the problems they engender. A right to property in traditional knowledge has been adopted as the basis of a compensatory justice approach to bioprospecting and biopiracy. Compensation may perpetuate unequal distributions when it is principally based on such contingent features as the location of biological resources and the existence of alternative potential sources. Institutions seeking to recognize the value of traditional knowledge and to establish a framework for benefit sharing need to endorse a new method of distribution. The difficult task of deciding how to distribute equitably is not tackled here but it should be pointed out that developing countries with valuable genetic resources stand a better chance of negotiating in their favor within the WTO framework than to seek compensation once resources have been exploited. It is also the case that a compensatory scheme perpetuates a proprietary conception of knowledge that may have the unintended consequence of strengthening arguments in favor of patent rights over such important goods as essential medicines.

REFERENCES

Adams WA (2002) Intellectual property infringement in global networks: the implications of protection ahead of the curve. Int'l J L & Info Tech 10(1):71–131

Ad hoc open-ended working group on access and benefit-sharing (2003) Report of the ad hoc open-ended working group on access and benefit-sharing on the work of its second meeting. Online: Convention on Biological Diversity <http://www.biodiv.org/doc/meetings/cop/cop-07/official/cop-07-06-en.pdf>

Beitz C (1978) Political theory and international relations. Princeton University Press, Princeton
Chapman AR (2002) The human rights implications of intellectual property protection. J Int'l Econ L 5(4):861–882
Davidson D (1984) On the very idea of a conceptual scheme. In: Davidson D (ed) Inquiries into truth and interpretation. OUP, Oxford
de Ortuzar MG (2003) Towards a universal definition of 'Benefit-Sharing'. In: Knoppers BM (ed) Populations and genetics: legal and socio-ethical perspectives. Martinus Nijhoff Publishers, Boston, pp 473–485
Downes D (1997) Using intellectual property as a tool to protect traditional knowledge: recommendations for the next steps. Centre for International Environmental Law, Washington
Downes DR (2000) How intellectual property could be a tool to protect traditional knowledge. Columbia J Envtl L 25(2):253–281
Downes DR, Laird SA (1999) Innovative mechanisms for sharing benefits of biodiversity and related knowledge case studies on geographical indications and trademarks. Prepared for UNCTAD Biotrade Initiative, available online: <http://www.ciel.org/Publications/InnovativeMechanisms.pdf>
Drahos P (1996) A philosophy of intellectual property. Dartmouth Pub. Co., Hanover
Drahos P (2001) Indigenous knowledge, intellectual property and biopiracy: is a global bio-collecting society the answer? European Intellectual Property Review Eur Intellect Property Rev 6:245–50
Gold ER, Caulfield TA (2003) Human genetic inventions, patenting and human rights. Report Commissioned by the Department of Justice, Government of Canada, p 46
Harding SG (1986) Science questions in feminism. Cornell University Press, Ithaca
Hettinger EC (1989) Justifying intellectual property. Phil Pub Affairs 18(1):31–52
Hughes J (1988) The philosophy of intellectual property. Geo L J 77(2):287–366
Human Genome Organisation (HUGO) (2000) Statement on benefit sharing. Vancouver, available online: HUGO <http://www.gene.ucl.ac.uk/hugo/benefit.html>
Juma C (1989) The gene hunter: biotechnology and the scramble for seeds. Princeton University Press, Princeton
Leskien D, Flitner M (June 1997) Intellectual property rights and plant genetic resources: options for a *Sui Generis* system. Issues in Genetic Resources 6:42
Macpherson CB (1978) The meaning of property. In: Macpherson CB (ed) Property: mainstream and critical positions. University of Toronto Press, Toronto
Nijar GS (1996) In defence of indigenous knowledge and biodiversity: a conceptual framework and essential elements of a rights regime. Third World Network, Penang
Pogge T (1994) An egalitarian law of peoples. Phil Pub Affairs 23(3):195–224
Rawls J (1999a) A theory of justice. Harvard University Press, Cambridge, MA
Rawls J (1999b) Law of peoples. Harvard University Press, Cambridge, MA
Sarma L (1999) Biopiracy: twentieth century imperialism in the form of international agreements. Temp Int'l & Comp L J 13(1):107–136
Secretariat of the Convention on Biological Diversity (2004) Access and Benefit-Sharing as Related to Genetic Resources. Decision VI/2, online: <http://www.biodiv.org/decisions/default.aspx?m=cop-06&d=24>
Shiva V (1997) Biopiracy: the plunder of nature and knowledge. South End Press, Boston
Shiva V (2000) Stolen harvest: the hijacking of the global food supply. South End Press, Cambridge, MA
Tobin B (1997) Know-how licenses: recognising indigenous rights over collective knowledge. Bulletin of the Working Group on Traditional Resource Rights 4:17–18
Trosow SE (2003) Fast-track trade authority and the free trade agreements: implications for copyright law. Can J L & Tech 2(2):135–149
UN Commission on Human Rights, High Commissioner (2001) The impact of the Agreement on Trade-Related Aspects of Intellectual Property Rights on Human Rights. Geneva, at par 41.
Walterscheid EC (1994) The early evolution of the U.S. Patent Law: Antecedents (Part 2). Journal of the Patent and Trademark Office Society 76(11):849–880

World International Property Office (WIPO) (2002) Intergovernmental committee on intellectual property and genetic resources, traditional knowledge and folklore (May 20, 2002), Operational terms and definitions. Geneva, WIPO/GRTKF/IC/3/9, online: <http://www.wipo.int/eng/meetings/2002/igc/pdf/grtkfic3_9.pdf> par. 25

DONNA CRAIG

CHAPTER 5

BIOLOGICAL RESOURCES, INTELLECTUAL PROPERTY RIGHTS AND INTERNATIONAL HUMAN RIGHTS: IMPACTS ON INDIGENOUS AND LOCAL COMMUNITIES

INTRODUCTION

"Some 200 million Indigenous people (4 percent of the world population) live in environments ranging from polar ice and snow to tropical deserts and rain forests. They are distinct cultural communities with land and other rights based on historical use and occupancy. Their cultures, economies and identities are inextricably tied to their traditional lands and resources. Hunting, fishing, trapping, gathering, herding or cultivation continue to be carried out for subsistence—food and materials—as well as for income" (World Conservation Union, 2001).

Indigenous Peoples also provide valuable resources and knowledge, about plant and animal use, including methods of preparation, storage and management, which is of global economic significance. Their biogenetic resources already form the basis of sizeable seed, pharmaceutical and natural product industries. Natural resource management, soil fertility maintenance, stream and coastal conservation and forest and agricultural system models provide viable, time-tested options for sustainable development adapted to microclimate variations and local socio-political ecosystems.

Yet Indigenous Peoples confront increasing external pressures on their lands, territories, resources, knowledge, innovations and practices. Even the recognition of their past endowments to world food and medicinal sources, as well as their significant contributions to agriculture, water and forest management, has done little to offset their political marginalisation.

ROLE OF TRADITIONAL ECOLOGICAL KNOWLEDGE (TEK) IN CONSERVING BIODIVERSITY

Traditional Ecological Knowledge (TEK), as defined by Johnson, covers "a body of knowledge built by a group of people through generations living in close contact with nature. It includes a system of classification, a set of empirical observations about the local environment, and a system of self management that governs resource use" (Johnson, 1992).

TEK not only refers to Indigenous communities, but also incorporates the knowledge base of hundreds of millions of members of predominately rural societies who depend on the natural environment for their livelihood. Their knowledge has been passed on over many thousands of years, accumulating intricate, detailed and sacred understandings of the local land to form their cultures and knowledge systems (IUCN, 1997). Traditional communities relationship with nature is such that the concepts of biodiversity and conservation are not foreign to them. Instead it is an "integral part of human existence, in which utilisation is part of the celebration of life" (Posey, 1999). These include notions of stewardship and totems that are fundamental to biodiversity conservation.

It is vital to incorporate TEK into international law and policy, as it will assist preservation of local knowledge, encourage participation by these communities in environmental management and (with prior informed consent) provide a framework for access and equitable benefit sharing. Age-old practices are not only beneficial to these communities but also provide examples of appropriate sustainable conservation strategies displaying efficient management techniques for soil, water, fisheries and forestries, which have wider applications.

OWNERSHIP AND PROTECTION OF TEK

Ownership and property are Western legal concepts that do not easily transpose into traditional and Indigenous systems. The notion of transferability is the least compatible element of property. This is because traditional and Indigenous Peoples identify themselves within communities who are tied to the land, often based on spiritual connections. There is often no single identifiable individual that could stand as the property owner of lands and biological resources.

Hansen and VanFleet (2003), argue that traditional knowledge is usually collective in nature and is often considered to be the property of the community as a whole:

> It is transmitted through specific cultural and traditional information exchange mechanisms, for example, maintained and transmitted orally through elders and specialists (breeders, healers, etc.), and often to only a select few people within a community.

The rich and complex systems of traditional knowledge protected and transmitted through customary law regimes are often undermined by dominant legal systems. Attempts by Indigenous Peoples to assert and protect their own approaches to TEK, in an increasingly globalised world, has often taken the form of a debate about the appropriateness, use and abuse of intellectual and cultural property rights. These issues will be canvassed briefly as they are dealt with in more detail by Onwuekwe and Mgbeoji respectively in Chapters 2 and 6 of this book.

WHO ARE INDIGENOUS PEOPLES?

"There are approximately 15,000 culturally-distinct ethnic communities in the world today (UNEP, 1992) and, while the diversity to be found among these cultures is both marvellous and extraordinary, most Indigenous Peoples share a sense of communal responsibility for their land and its living resources" (Rural Advancement Foundational International, 1994).

The definitions of Indigenous Peoples are varied and problematic in some parts of the world. The Special Rapporteur of the United Nations Economic and Social Council Sub-Commission on the Prevention of Discrimination of Minorities defined Indigenous Peoples in the following manner:

> Indigenous communities, Peoples and nations are those which, having a historical continuity with pre-invasion and pre-colonial societies that have developed on their territories, consider themselves distinct from other sectors of the societies now prevailing in those territories, or parts of them. They form at present non-dominant sectors of society and are determined to preserve, develop and transmit to future generations their ancestral territories, and their ethnic identity, as the basis of their continued existence as Peoples, in accordance with their own cultural patterns, social institutions and legal systems (UN, 1986).

The International Labour Organisation (ILO) Convention 169 "Concerning Indigenous Peoples in Independent Countries" (1989), identifies Indigenous Peoples as:
a) Tribal Peoples in countries whose social, cultural and economic conditions distinguish them from other sections of the national community, and whose status is regulated wholly or partially by their own customs or traditions or by special laws or regulations, and
b) Peoples in countries who are regarded by themselves or others as Indigenous on account of their descent from the populations that inhabited the country, or a geographical region to which the country belong, at the time of conquest or colonization or the establishment of present state boundaries and who, irrespective of all their legal status, retain, or wish to retain, some or all of their own social, economic, spiritual, cultural and political characteristics and institutions.

The fundamental principle established by ILO 169 is that self-identification, "as Indigenous or tribal shall be regarded as a fundamental criterion for determining the groups to which the provisions of this convention apply" (Posey, 1999).

WHAT ARE INDIGENOUS INTELLECTUAL AND CULTURAL PROPERTY RIGHTS?

Intellectual and cultural property refers to Indigenous knowledge, whether it is in the form of biological knowledge, customary knowledge or created tangible materials, that is passed on from one generation to the next.

"A Study on the Protection of Cultural and Intellectual Property of Indigenous Peoples" by Daes argues that the separation of "intellectual" and "cultural" property is inappropriate. Instead, a holistic and integrated view of Indigenous heritage,

encompassing all aspects of their lives, is essential. This is a more compatible view as it reflects Indigenous philosophies of oneness with their Peoples and lands:

> ...heritage includes all expressions of the relationship between the people, their land and the other living beings and spirits which share the land, and is the basis for maintaining social, economic and diplomatic relationships—through sharing—with other Peoples. All of the aspects of heritage are interrelated and cannot be separated from the traditional Territory of the people concerned. What tangible and intangible items constitute the heritage of a particular Indigenous Peoples' must be decided by the people themselves...(Daes, 1993).

In formulating the concept of property rights to protect Indigenous culture, it is important to recognise the needs and rights of Indigenous Peoples' under international and domestic law. In failing to do so, we will have a system of property rights that does not adequately address their needs and furthermore may work to disintegrate their community.

The following is a list of some key heritage rights that Indigenous Peoples in Australia call for in relation to intellectual and cultural property:

- The right to own and control Indigenous intellectual and cultural property,
- The right to define what constitutes Indigenous intellectual and cultural property and/or Indigenous heritage,
- The right to ensure that any means of protecting Indigenous intellectual and cultural property is premised on the principle of self-determination, which includes the right and duty of Indigenous Peoples to maintain and develop their own cultures and knowledge systems and forms of social organisation,
- The right to authorise, or to refuse to authorise, the commercial use of Indigenous intellectual and cultural property in accordance with customary law,
- The right to benefit commercially from the authorised use of Indigenous intellectual and cultural property, including the right to negotiate terms of such usage,
- The right to protect Indigenous sites, including sacred sites,
- The right to control the disclosure, dissemination, reproduction and recording of Indigenous knowledge, ideas and innovations concerning medicinal plants, biodiversity and environmental management,
- The right to own and control management of land and sea, conserved in whole or part because of their Indigenous cultural values (Janke and Frankel, 1998).

The international law for the protection of intellectual property has been developed with virtually no regard for the needs of Indigenous Peoples for the protection of their cultural and intellectual property (usually embodied in their communal lifestyles). They were developed to protect the marketable property of non-Indigenous corporations. This is a restricted form of property which is severed from the original components of the "invention" and the cultures which may have nurtured its initial stages. It is the "modification" or "discovery" through non-Indigenous technology which is usually rewarded and protected by intellectual property rights.

Indigenous Peoples, in contemporary society, require an economic base. This is even the case where many of their activities and lifestyle remain "traditional." In many situations their ecosystems have been altered, and political circumstances changed, so that they cannot (or do not wish) to live a totally subsistence lifestyle. The economic and sustainable use of resources may be consistent with the maintenance of Indigenous lifestyle or cultural adaptation over time. Intellectual property rights are one type of legal strategy that is being considered by Indigenous Peoples, to protect their biological resources, cultural practices and to develop an economic base required by contemporary circumstances.

It is difficult to formulate a version of intellectual and cultural property rights appropriate for Indigenous Peoples at the present time. Some of the difficulties with existing laws are as follows:

- Indigenous Peoples are not given recognition, as legal "persons," to enforce international conventions on intellectual property.
- There is little experience with identifying the Indigenous knowledge or resource "component," when it is modified by industrial users and with placing an economic value on such a "component." Such knowledge is usually part of a culture that cannot be segmented by Indigenous Peoples or others.
- Intellectual property laws require an individual group, or other legal entity, to make the claim. Indigenous knowledge is often communally held (Yamin and Posey, 1993). This is not an impediment in itself, if groups who control knowledge and resources can be identified. The boundaries of groups and the "exclusivity" of knowledge and practice may be problematic.
- Intellectual property rights, in the form held by Indigenous Peoples, may not be readily marketable. Large scale marketing may also impact on the cultures and lifestyles which produced the knowledge and technologies.
- Indigenous Peoples would require greater resources and capacity building to avail themselves of intellectual property laws (Yamin and Posey, 1993).

The intellectual and cultural property rights of Indigenous Peoples require further consideration and development having regard to the specific provisions of the 1992 UN *Convention on Biological Diversity* (discussed in a later section of this chapter).

The emphasis in this chapter will be on important international legal standards relating to human and environment rights of Indigenous Peoples to their biological resources. The difficulties in developing the most appropriate forms of intellectual property rights for the recognition of Indigenous rights and aspirations does not mean that there is a legal void. There is a rich and diverse range of applicable international legal norms. The challenge is to find practical and enforceable means to implement them in the international and domestic context. Indigenous Peoples' have their own distinct views on how this should be done and their approaches require much greater research, support and consideration by the international community and nation States.

SUSTAINABLE DEVELOPMENT STRATEGIES

Indigenous Peoples are inevitably impacted by decisions and plans which go beyond their land and involve Peoples from the dominant culture. A key issue which must be addressed is how we recognise the requirements of Sustainable Development (SD) and the needs of Indigenous Peoples' for self-determination (including their own development strategies). It is fundamental that Indigenous Peoples must have legally recognised title to their land, seas and natural resources and the power to control their use and management in a way that they consider appropriate.

Following the United Nations Conference on the Human Environment (United Nations, 1972), several declarations have stated the basic right of Indigenous Peoples to exercise self-determination to achieve SD:

> Development should respect, maintain and enhance the diversity of natural life and human culture to maintain and expand the availability of options for this and future generations .... This requires that homogenisation of land use and human lifestyles be avoided (Cocoyoc Declaration, 1974).

The *Declaration of San Jose* defines the right to ethno-development as:

> [The] amplification and consolidation of... a culturally distinct society's own culture, through the strengthening of its capacity to guide its own development and exercise self-determination ... and implying an equitable and proper organisation of power (Declaration of San Jose UNESCO, 1981).

This chapter explores evolving standards and legal approaches to human, environmental and Indigenous rights and the extent to which they potentially contribute to the SD. It will be argued that Indigenous intellectual and cultural property rights need to be understood, and recognised, in the context of these rights under international law.

STANDARDS AND APPROACHES TO HUMAN AND ENVIRONMENTAL RIGHTS OF INDIGENOUS PEOPLE

Traditionally, international law has focused on state sovereignty and nation states have predominated as legitimate parties. This has changed to a limited extent, in regimes such as the human rights conventions, which focus on the rights of the individuals. The role of Non Governmental Organizations (NGO's) in the negotiation and implementation of some global and regional environmental conventions has also challenged the dominant concept that international law purely regulates the relationship between nation states. These new challenges, in modern international law, relate to the:

- concept and formulation of the rights (including substantive and procedural rights);
- need to address the collective rights of Peoples (particularly raised in *ILO 169* and the *Draft UN Declaration on the Rights of Peoples*).
- enforcement, monitoring and financing of the obligations of developing nations on an equitable basis (sometimes involving differential responsibilities for the "North" and "South");
- need to develop practical, affordable and innovative implementation strategies that empower, legitimise rights (in terms of the dominant political order), work effectively for local communities and Indigenous and tribal Peoples and address urgent needs such as poverty and;
- promotion of the rapid dissemination and adaptation of comparative experiences relating to national implementation.

Often, Indigenous Peoples are seeking both collective and individual rights. The predominant western legal traditions, based on social contract theory, tends to put individual rights against the state (Clinton, 1990). Indigenous strategies often involve this defensive role but they also assert their collective rights and (internal) collective obligations to maintain and continue to evolve their cultures and nations in the face of threats such as discrimination, dispossession, environmental change and the cultural, economic and the impacts of the dominant society.

Because Indigenous Peoples have integral and unique relationships with the earth (including land, seas, resources, wildlife) they do not fragment or compartmentalise their rights and obligations relating to their ecological, spiritual, cultural, economic and social dimensions. The accumulated knowledge acquired in this way, has become a matter of global interest and exploitation (Posey, 1999). This can also be seen as a fundamental illustration of SD as it draws "wisdom" from the idea of integrating these dimensions. SD has now become the central focus of modern environmental policy and rapidly evolving human rights, such as the right to development and the right to a healthy environment, (see national constitutions with rights to environment quality such as Philippines, Pakistan, and India).

Human rights are often characterised as follows: civil and political rights; economic, social and cultural rights; and the (evolving) right to development and the right to a healthy environment.

Article 1 of the *Declaration on the Right to Development* (GA Res 128, 1986) reads:

> The right to development is an inalienable human right by virtue of which every human person and all Peoples are entitled to participate in, contribute to and enjoy economic, social, cultural and political development, in which all human rights and fundamental freedom can be fully realised.

International law and policy relating to SD has been promoted by *Rio Declaration*, *Agenda 21*, *Convention on Biological Diversity* (CBD), *Framework Convention on Climate Change* (FCCC) and the *Forest Principles* which were the key outcomes of the *United Nations Conference on Environment and Development* (UNCED) in 1992 (Craig and Ponce, 1995). The rights of Indigenous Peoples' are specifically mentioned in each of these global declarations and instruments except for the FCCC. The *Plan of Implementation* developed at the *World Summit on Sustainable Development* (held in 1998 at Johannesburg, South Africa.) contains some significant soft law provisions relating to Indigenous rights.

Arguably, basic provisions in international human rights instruments have become part of international customary law, known as *"jus cogens,"* and they are inextricably related to the rights to environment and development. Important *jus cogens* norms are as follows:

- the right to be free from genocide (including cultural genocide)
- the right to life (including the right of people not to be deprived of their own means of subsistence)
- the right of Peoples to self-determination (World Summit on Sustainable Development, 2002)

Ward(1993) goes on to argue that these provisions in human rights instruments (now *jus cogens*) establish:

> [T]he right to be free from hunger, the right to an adequate standard of living, the right to health, free personal development, sustainable community development using environmental and social impact assessments, and the right to integrity of the community as a whole.

THE INTERNATIONAL COVENANT ON ECONOMIC, SOCIAL AND CULTURAL RIGHTS

After the adoption of the *Universal Declaration of Human Rights* (UDHR), the United Nations' Commission on Human Rights commenced drafting two human rights' conventions which are relevant to Indigenous Peoples' rights to ecologically SD.

Article 1(2) of the International Covenant for Economic, Social and Cultural Rights (ICESCR) provides that:

> [A]ll Peoples may, for their own ends, freely dispose of their natural wealth and resources without prejudice to any obligations arising out of international economic co-operation, based upon the principle of mutual benefit and international law. In no case may a people be deprived of its own means of subsistence.

Indigenous Peoples assert their rights as "people" under this provision, but prefer the terminology of Peoples associated with the right to self-determination.

Article 15(1c) provides that States recognise the right to "benefit from the protection of the moral and material interests resulting from any scientific, literary or artistic production of which he is the author".

INTERNATIONAL COVENANT ON CIVIL AND POLITICAL RIGHTS

Article 1(2) of the ICCPR is identical to Article 1(2) of the ICESCR. Article 18 of the ICCPR states that everyone shall have a right to freedom of thought, conscience and religion. Article 27 provides for the rights of minorities:

> In those States in which ethnic, religious or linguistic minorities exist, persons belonging to such minorities shall not be denied the right, in community with other members of their group, to enjoy their own culture, to profess and practise their own religion, or to use their own language.

HUMAN RIGHTS RELATING TO INTELLECTUAL PROPERTY IN BIOLOGICAL RESOURCE

Hansen and VanFleet argue that:

> Since the adoption of the Universal Declaration of Human Rights (UDHR) in 1948, intellectual property (IP) has been considered a fundamental human right of all Peoples. Article 27 of the Declaration states that everyone has the "right to protection of the moral and material interests resulting from any scientific, literary or artistic production of which he is the author." Since 1948, many international human rights instruments and documents have reinforced the importance of IP as a human right (Hansen and VanFleet, 2003).

Specific consideration needs to be given to Article 27 of UDHR, which also states, "Everyone has the right freely to participate in the cultural life of the community, to enjoy the arts and to share in scientific advancement and its benefits." Other international human rights instruments addressing intellectual property include the *International Convenant on Economic, Social and Cultural Rights* (ICESCR) Article 15, *International Labour Organization Convention No. 169* (ILO 169) Article 15, *Convention on Biological Diversity* (CBD) Article 8(j) and the *Draft Declaration on Indigenous Rights*, Article 29 (Hansen and VanFleet, 2003).

SPECIFIC INTERNATIONAL LEGAL INSTRUMENTS RELATING TO THE RIGHTS OF INDIGENOUS PEOPLES

The international legal status of Indigenous Peoples, as self determining nations, remains an enduring and controversial issue in international law. This situation continues even though there is increasing domestic legal recognition of various forms of Indigenous self government within many nations. The fear of succession, and other political factors, has influenced the international position adopted by States. The right of self determination remains a central concern in the negotiation of the UN *Draft Declaration of Indigenous Rights*. It is clear that this right is the focal point for asserting contemporary Indigenous claims for their integrated and comprehensive rights.

ILO CONVENTION 169

In 1989 the International Labour Organisation adopted a new convention, *ILO 169, Concerning Indigenous and Tribal Peoples in Independent Countries*. Some Indigenous Peoples and other commentators consider that *ILO 169* undermines Indigenous aspirations by emphasising "participation" or "consultation" rather than self determination. However, taken as a whole, ILO 169 is the most significant international treaty, so far, that recognises the integrated and comprehensive rights of Indigenous and Tribal Peoples.

Relevant articles require systematic and co-ordinated action to protect the integrated rights of Indigenous Peoples:

- 7.1 illustrates some elements of self-determination: "The Peoples concerned shall have the right to decide their own priorities..." but then is qualified, "for the process of development."
- 7.2 and 7.3 refer to improved living conditions with "Indigenous participation and co-operation." The carrying out of studies, although only "where appropriate", should include assessment of "social, spiritual, cultural and environmental impact."
- 13.1 and 13.2 incorporate concepts of Indigenous rights, requiring governments to "respect the special importance for the cultures and spiritual values of the Peoples" and their territories (which covers the total environment of areas which the peoples concerned occupy or otherwise use).
- 14.1 is a recognition of Indigenous rights to "ownership and possession... over the lands which they traditionally occupy." Furthermore, this provision extends to give rights to Peoples who do not have exclusive use of certain lands, but traditionally require access to it, such as "nomadic Peoples and shifting cultivators."
- 14.2 and 14.3 establishes the responsibility of governments to "guarantee effective protection of their rights of ownership and possession" and provide adequate legal mechanisms to dissolve possible land claims.
- Article 15.1 states, "The rights of the Peoples concerned to the natural resources pertaining to their lands shall be especially safeguarded. These rights include the right of these Peoples to participate in the use, management and conservation of these resources."

ILO 169 tends to be regarded as a "floor" and not the "upper level" of Indigenous rights. It was negotiated at the same time as the *Draft UN Declaration of Indigenous Rights*. Many Indigenous Peoples prefer the stronger language in the *Draft Declaration*. ILO is expressed in a form to encourage ratification by the maximum number of nations, some of which give poor recognition to Indigenous rights at the present time. International standards will develop under this convention along with other human rights and environmental conventions as well as "soft law" such as Universal Declarations by the United Nations.

DRAFT UNIVERSAL DECLARATION ON THE RIGHTS OF INDIGENOUS PEOPLES

In 1982 the Economic and Social Council of the United Nations authorised the establishment of a Working Group on Indigenous Populations to meet annually as part of the Sub-Commission on the Prevention of Discrimination and Protection of Minorities. The Mandate of the Working Group was:
- to review developments pertaining to the promotion and protection of the human rights and fundamental freedoms of Indigenous Peoples; and
- to give special attention to the evolution of standards concerning the rights of Indigenous populations.

The two major matters for Working Group deliberations are a "Review of Developments" and "Standard Setting". Since 1985 the Working Group has decided to meet its "Standard Setting" mandate by formulating a *Draft Declaration on the Rights of Indigenous Peoples*. The process has involved hundreds of Indigenous Peoples in UN meetings and work for over ten years. The *Draft Declaration* has been passed up to the Sub-Commission and it eventually will be considered by the UN General Assembly. Within the next few years it will be a UN Declaration with the same status as the *UN Declaration on Human Rights*. The Working Group Draft will be amended by Nation States in this process.

However, the Working Group version of the *Draft Declaration* (1993) is the most comprehensive, integrated and strongest articulation of Indigenous rights, by Indigenous Peoples, to date. This version expresses the international standard developed by Indigenous participants over a very long period of deliberation. This was one of the most participatory NGO processes, particularly including Indigenous Peoples, in UN history. The *Draft Declaration* needs to be read holistically as the clearest articulation of the integrated rights of Indigenous Peoples in evolving international law and policy.

The clear and unqualified right to Indigenous self determination is contained in Article 3. Other rights (including management rights) to lands, resources, waters, seas, biological resources, the recognition of intellectual and cultural property and the rights of Indigenous Peoples to determine their own development priorities are contained in Articles 25–30. These are more powerful reflections of the contemporary aspirations and needs of Indigenous Peoples than the provisions in ILO 169.

The *Draft Declaration* is slowly proceeding through the UN process. It will remain important to consider what parts of it are reflected in other treaties (discussed above), international customary law and *jus cogens*.

EVOLVING APPROACHES TO HUMAN RIGHTS AND THE ENVIRONMENT

The overlaps in the approaches to human rights has caused the United Nations to begin preparing a *Declaration of Principles on Human Rights and the Environment* (Meeting of Experts on Human Rights and Environment, 1994). This *Draft Declaration* commences as follows:

1. Human rights, an ecologically sound environment, sustainable development and peace are interdependent and indivisible.
2. All persons have the right to a secure, healthy and ecologically sound environment. This right and other human rights, including civil, cultural, economic, political and social rights are universal.

The World Conservation Union (IUCN) has prepared a Draft of an umbrella global treaty to provide a legal framework for SD. It is known as the *Draft Covenant on Environment and Development* (IUCN, 2000). This is a very important indicator of future trends in international environmental law. It includes respect for all life forms. It recognises the right to development but has regard to the urgent need to maintain and restore the earth's ecological systems.

The United Nations *Meeting of Experts on Human Rights and the Environment* (2002) has reviewed the progress made since the United Nations Conference on Environment and Development held in 1992 at Rio de Janeiro, Brazil. They concluded that there was a growing and close connection between human rights and environmental protection in the context of SD. Important developments have occurred at the national and international level. The experts particularly noted that the linkage is reflected in recent developments relating to procedural and substantive rights, in the activities of international organisations and in the drafting and application of national constitutions (United Nations, 2002).

There is a wealth of case law, particularly in developing nations, upholding the constitutional right to environmental quality (Craig et al., 2002). This often required innovative approaches to standing and public participation as seen in the famous Phillipines case of *Oposa* v. *Factoran* (GR No. 101083, 1993) and in the *Aarhus Convention on Access to Information, Public Participation and Access to Justice in Environmental Matters* (1998). The decisions of international treaty bodies (including courts and commissions) often recognise the violation of a fundamental human right as the cause, or result, of environmental degradation (GR No. 101083, 1993).

Some human rights treaties, such as the *Convention on the Rights of the Child* and *ILO 169 Concerning Indigenous and Tribal Peoples in Independent Countries*, the *African Charter on Human and Peoples Rights* and the *Protocol of San Salvador to the American Convention on Human Rights* expressly recognise the right to live in a healthy or satisfactory Environment (GR No. 101083, 1993). Respect for human rights is broadly accepted as a pre-condition for SD and that environmental protection is a pre-condition for the enjoyment of human rights—they are interdependent and interrelated (GR No. 01083, 1993). This is now reflected in national and international practices and developments (GR No. 101083, 1993).

It is also understood that poverty is at the centre of a number of human rights violations and is at the same time a major obstacle to achieving SD (GR No. 101083, 1993).

It no longer makes any sense to talk about "categories of rights". None of the human rights instruments attempt any ranking—all rights and freedoms are placed on an equal footing. "All rights are, like humans themselves, inextricably singular and social" (Shutkin, 1991). Therefore, "first" and "second" generation should be approached as a unity rather than a duality (Shutkin, 1991). Brownlie (1992) maintains that there has been an assumption lying behind the classical formulation of human rights standards that "group rights would be taken care of automatically as a result of the protection of the rights of individuals." This approach has been inadequate to protect the rights of Indigenous Peoples and there is an urgent need to advance international law in the direction of a more unified conceptual framework.

International Labour Organisation Convention 169 (ILO 169) and the *Draft Declaration on the Rights of Indigenous Peoples* (UN Commission on Human Rights, Sub-Commission on Prevention of Discrimination and Protection of Minorities) provide examples of the integrated rights approach (the *Draft Declaration* being a better example). ILO 169 and *Draft Declaration* should be understood within the wider framework of international environmental, Indigenous rights and human rights law. This is discussed by Posey who argued that there has been a gradual evolution of *sui generis traditional resource rights* (Posey, 1996). Developing new forms of intellectual and cultural property rights should be a part of this process (not an impediment of it).

Thus, Indigenous rights in international law and policy are not the "poor relation." They are a catalyst for developing more unified and effective approaches to human rights and sustainable development which should be recognised and implemented.

AN EXAMPLE OF INTEGRATED RIGHTS APPROACHES: TRADITIONAL RESOURCE RIGHTS AND *SUI GENERIS* SYSTEMS

Sui generis is the Latin word for "unique" or "of its own kind". In a legal context, the reference to a *sui generis* system involves an alterative or modified framework to the existing law, governed by fundamentally different principles and mechanisms.

Posey clarifies the definition and application of a *sui generis* approach in the discussion of Article 27.3(b) of the *Trade Related Aspects of Intellectual Property Rights Agreement* (TRIPs) (1994), which provides for the "...the protection of plant varieties either by patents or by an effective *sui generis* system or by any combination thereof." *Sui generis* systems in this context can mean a modification of existing intellectual property law, a new intellectual property right or an alternative to intellectual property rights.

The *sui generis* system referred to in TRIPs illustrates the acknowledgement that current intellectual property laws, specifically by way of patents, may be an inappropriate method of protection of Indigenous culture and heritage.

At a time when the difficulties in adapting existing western legal structures to enhance conservation of biodiversity and empower Indigenous and local communities become manifest, the emergence of the concept of Traditional Resource Rights (TRR) appears to be timely. The TRR concept was first elaborated in an article by Posey entitled: "Traditional Resource Rights: De Facto Self-determination for Indigenous Peoples," (TRIPs, 1994) and it has been proposed as a specific example of a *sui generis* system of legal rights.

The term TRR seeks to go beyond limited concepts of intellectual property rights and describes the many "bundles" of rights existing and being developed which can be utilised for the protection of and compensation for the use of traditional knowledge and resources. Traditional resources include plants, animals, and other material objects, like minerals and cultural artifacts, which may have intangible (e.g. sacred, ceremonial, heritage, or aesthetic) qualities. Traditional resources may also be totally intangible, metaphysical, or non-quantifiable with no physical manifestations, such as systems of knowledge. The term "property" is inappropriate, since property in traditional societies frequently has intangible, spiritual manifestations, and, although worthy of protection, can belong to no human being.

Traditional Resource Rights encompasses not only conventional intellectual property rights mechanisms but basic rights and customary law defined in, or accommodated by, national and international laws, agreements and declarations. The TRR concept can form the basis for *sui generis* systems for protection and benefit-sharing. Also, it could be implemented locally, nationally and internationally as a set of principles to guide the process for dialogue between Indigenous and local communities, and governmental and non-governmental institutions. TRR could, for example, guide the development of innovative contracts providing benefits to local communities in exchange for the transfer of information and biogenetic material.

In short, TRR-guided negotiations can offer challenging opportunities for new partnerships based on increased respect for Indigenous Peoples and their knowledge, new codes of ethics and standards of conduct, socially and ecologically responsible practices, and holistic approaches to sustainability. At the very least TRR-oriented discussions are certain to be more fruitful than those based on intellectual property rights because TRR brings the environmental concerns of local communities, and issues relating to SD and global trade, into the human rights debate. The process of developing TRR will involve further dialogue, debate and consciousness-raising.

Traditional Resource Rights consists of the following bundles of rights, which are supported by the legally and non-legally binding agreements outlined in Table 1, which appears on page 93.

Transfer of plant and animal genetic materials takes on a new dimension since such materials should appropriately be seen as technological transfers as they are the application of traditional knowledge and lifeways systems. Establishment of global funds, such as the one for "Farmers' Rights" to compensate local farmers for their "contributions in conserving, improving and making available plant genetic resources" (FAO Resolution 4/89), will become essential to effect equitable sharing of the benefits of biodiversity conservation. The provision of funding, loans

TABLE 1. Traditional Resource Rights

Right (bundle)	Supporting agreements: legally binding	Supporting agreements: non legally binding
Human rights	ICESCR, ICCPR, CDW, CERD, CG, CRC, NLs	UDHR, DDRIP, VDPA
Right to self-determination	ILO169, ICESCR, ICCPR	DDRIP, VDPA
Collective rights	ILO169, ICESCR, ICCPR	DDRIP, VDPA
Land and territorial rights	ILO169, NLs	DDRIP
Right to religious freedom	ICCPR, NLs	UDHR
Right to development	ICESCR, ICCPR, ILO169	DDHRE, DDRIP, DHRD, VDPA
The right to privacy	ICCPR, NLs	UDHR
Prior informed consent	CBD, NLs	DDRIP
Environmental integrity	CBD	RD, DDHRE
Intellectual property rights	CBD, WIPO, GATT, UPOV, NLs	
Neighbouring rights	RC, NLs	
Right to enter into legal agreements, such as contracts and covenants	NLs	
Cultural property rights	UNESCO-CCP, NLs	
Right to protection of folklore	NLs	UNESCO-WIPO, UNESCO-F
Right to protection of cultural heritage	UNESCO-WHC, NLs	UNESCO-PICC
Recognition of cultural landscape	UNESCO-WHC	
Recognition of customary law and practice	ILO169, NLs	DDRIP
Farmers' Rights		FAO-IUPGR

Source: Adapted from Posey, 1996 at p. xiv.

and foreign aid (by nations, multilateral lending institutions and other agencies), affecting and involving Indigenous Peoples, should require the development of TRR *sui generis* provisions.

COMPLIANCE WITH INTERNATIONAL STANDARDS

The enforcement of international law and policy, relating to the human and environmental rights of Indigenous Peoples, is often inadequate with the possible exception of the *Convention on the Rights of the Child, ILO 169* and the optional protocol under the *Convention on Civil and Political Rights*. Regional human rights conventions (Europe and the Americas) arguably have stronger enforcement and complaint mechanisms. Increasingly, remedies and settlements being sought by Indigenous Peoples involve co-management regimes. Many disputes, such as the *Awas Tingni* case discussed below, arise out of resource allocation and management decisions on Indigenous lands that are made without their involvement.

Inter-American Human Rights Regime

The Organization of American States (OAS) is a regional agreement, founded on the principle that States, comprising North, Central and South Americas, must unify their efforts to ensure personal liberty and social justice based on essential human rights. The OAS system promotes the protections of human rights by establishing substantive norms, supervisory institutions, and accessible petition procedures. Member States of the OAS must accept the human rights standards indicated in the *Charter of the Organization of American States* and the *American Convention on Human Rights* which are enforced by the Inter-American Commission on Human Rights.

The characteristic that distinguishes it from the United Nations system is that accessibility of the "communication" of a petition is not restricted to NGOs or member states but the individual, family relation or third party may initiate proceedings. The Commission can appoint moderators to oversee "friendly settlement" proceedings whereby the parties agree, acting in good faith, to attempt to reach a resolution of the violations and have a 6 month deadline. A restriction on the OAS system for the protection of human rights is that similar proceedings cannot be ongoing in the United Nations system concurrently. The finding of a violation of a member State and subsequent ruling of the Inter-American Commission is not legally binding on the parties. However, should the member State or Commission wish to pursue the matter further, they may petition the Inter-American Court of Human Rights for a hearing. The State must first accept the jurisdiction of the Court in order to make any decision legally binding on the parties and any non-compliance will be then enforceable in the domestic court system.

A recent decision of the Inter-American Court of Human Rights, *The Mayagna (Sumo) Indigenous Community of Awas Tingni v. The Republic of Nicaragua* (Awas Tingni), was a landmark decision of the Court that recognised duties of the State to demarcate and preserve the integrity of ancestral lands of Indigenous Peoples within its boundaries. It further recognised the right of all persons to expeditious and fair access to competent tribunals as not only a pillar of the American Convention but a Rule of Law in democratic societies.

The Court stated that it is not enough for Nicaragua to acknowledge the rights of Indigenous Peoples to their ancestral territories. It must put this recognition into practice by not engaging in actions that would diminish or affect Indigenous rights or interests in ancestral lands. This decision indicates that compliance may require an effective system of national implementation to give practical meaning to the legal rights of Indigenous Peoples. As compensation for moral damage, by allowing the harvesting of natural resources on traditional Awas Tingni land without notification, consultation or compensation, the Nicaraguan government was required to pay a monetary award of US $50,000.00.

United Nations Permanent Forum on Indigenous Peoples

The announcement of the United Nations Decade for the World's Indigenous Peoples was indicative that the recognition of the fundamental rights of Indigenous Peoples will continue to evolve internationally and nationally. The Permanent Forum on Indigenous Peoples, a new subordinate organ of the United Nations Economic and Social Council, has an innovative internal organization. It is comprised of eight representatives of Indigenous Peoples and eight experts chosen by the member states of the United Nations, a combination not previously seen in other United Nations entities. The Forum is in response to the growing body of work on Indigenous issues at the international level and the resounding success and participation of Working Groups[1] under the Human Rights Commission.

RECOGNITION OF INDIGENOUS RIGHTS IN INTERNATIONAL ENVIRONMENTAL LAW

World Heritage Convention

The *World Cultural and Natural Heritage Convention* (1972) provides for the identification and protection of cultural and natural heritage which is of "outstanding universal value." A World Heritage List is compiled under the auspices of UNESCO. The State parties to the Convention also provide some (limited) funds for the protection of the places on the list.

Examples of criteria for "cultural" properties require that they should:
- be an outstanding example of a traditional human settlement or landuse which is representative of a culture (or cultures), especially when it has become vulnerable under the impact of irreversible change; or,
- be directly or tangibly associated with events or living traditions, with ideas, or with beliefs, with artistic and literary works of outstanding universal significance. (Note that the Committee considers that this criterion should justify inclusion in the list only in exceptional circumstances or in conjunction with other criteria).

The criteria for "natural" properties includes places which:
- be outstanding examples representing major stages of earth's history, including the record of life, significant on-going geological processes in the development of landforms, or significant geomorphic or physiographic features; or
- be outstanding examples representing significant on-going ecological and biological processes in the evolution and development of terrestrial, fresh water, coastal and marine ecosystems and communities of plants and animals; or
- contain superlative natural phenomena or areas of exceptional natural beauty and aesthetic importance; or
- contain the most important and significant natural habitats for in-situ conservation of biological diversity, including those containing threatened species of outstanding universal value from the point of view of science or conservation;

Of the 754 properties listed sites (as at 3 July 2003), 582 were selected for their cultural importance and 149 for their natural significance. The rest are joint cultural and natural heritage sites.

The World Heritage Site of Tongariro National Park in New Zealand was selected due to the area's importance in Maori mythology and the sacred nature of the mountains. The World Heritage Committee concluded after considering its inclusion under the criteria that it was: an outstanding example of an associative cultural landscape tied to the cultural identity of the Maori people.[2]

It would appear from the above that the Convention may be useful in protecting the cultural heritage of some Indigenous Peoples although recognition of their concerns and values, through criteria, has been slow. A factor will be the willingness of the World Heritage Committee and "expert" advisers such as the World Conservation Union to pay heed to the interests of Indigenous Peoples when considering the inclusion of new nominations and already listed properties under the new criteria. However, ultimately, the extent to which the religious and cultural importance of places and objects for ethnic minorities and Indigenous Peoples is taken into account in the World Heritage List depends upon: (a) whether governments are willing to consult Indigenous Peoples: (b) whether national legislation to implement the Convention allows for a flexible or broad interpretation of "cultural and national heritage" and (c) whether the Committee is prepared to take the view that cultural and natural properties important to an Indigenous people constitute part of the heritage of humankind of sufficient importance to justify the expense of their protection (Posey, 1999). Precedents do exist and some places that are important to an Indigenous people are now on the World Heritage List, such as Tongariro and Uluru Kata Tjuta (Ayers Rock) in Australia, which are sacred to Indigenous Peoples.

Convention to Safeguard Intangible Cultural Heritage

The intangible nature of much of Indigenous knowledge and culture has been an overriding cause for the incompatibilities that these communities face with intellectual property. By the early 1980s the awareness of the concept of intangible expressions of spirituality emerged (UNESCO, 2003). In October 2003, by overwhelming majority, the member States of UNESCO adopted the *International Convention to Safeguard Intangible Cultural Heritage*.

The primary aims are to assist in the preservation of intangible heritage and customs, raise awareness at all levels and to integrate international participation. Specific requirements such as the drawing up of national inventories of cultural property to be protected, funding provided by States parties to UNESCO and the establishment of an Intergovernmental Committee.

This Convention, possibly a few years away from international implementation, illustrates the progressive developments towards Indigenous heritage and cultural protection. Further, it highlights the inadequacies of the current legal attempts of providing protection through international intellectual property law.

UN Convention on Biological Diversity

The *Convention on Biological Diversity* (CBD) seeks to guide and govern the use and conservation of biogenetic resources and traditional knowledge, while protecting local communities and Indigenous Peoples. The primary role and sovereignty of States is not challenged by the CBD. For example, Article 1 of the Convention states:

> The objectives of this Convention ... are the conservation of biological diversity, the sustainable use of its components and the fair and equitable sharing of the benefits arising out of the utilisation of genetic resources, including by appropriate access to genetic resources and appropriate transfer of relevant technologies, taking into account all rights over those resources and to technologies, and by appropriate funding.

The CBD places Indigenous and traditional knowledge, as well as traditional technologies and biogenetic resources, under Nation-State sovereignty. Thus, no matter how liberal or generous provisions might appear, Indigenous Peoples are faced with a difficult conundrum. On the one hand their contributions, central role in SD and conservation, and rights as decision-makers and beneficiaries are recognised far beyond any previous international binding legal convention. However, Indigenous Peoples are reluctant to accept that ultimate control over resources lies with nation-states. Few Indigenous groups are willing to allow this *a priori* usurpation of their fundamental rights of self-determination no matter what promises and favourable interpretations may arise.

There is no reason to expect that the CBD will significantly contribute to the resolution of basic issues raised in the *Draft UN Declaration on the Rights of Indigenous Peoples*, namely their call for self-determination. However, it may help pave the way for the development of useful instruments that work towards more equitable partnerships with Indigenous Peoples.

Intellectual and cultural property rights and TRR's are clearly mechanisms for renegotiating the terms of these partnerships, which would be built upon the recognition, application, and compensation for Indigenous technologies and knowledge. Unfortunately, the CBD does not provide any explicit legal means to recognise, protect or compensate Indigenous Peoples, nor do such mechanisms exist in any other global legal forum.

The CBD itself refers to Indigenous Peoples in several significant sections. Article 8(j) is the most important of these sections:

> Subject to it's national legislation, respect, preserve and maintain knowledge, innovations, and practices of Indigenous and local communities embodying traditional lifestyles relevant for the conservation and sustainable use of biological diversity and promote their wider application with the approval and involvement of the holders of such knowledge, innovations and practices and encourage the equitable sharing of the benefits arising from the utilisation of such knowledge, innovations and practices.

The Explanatory Guide to the CBD (IUCN, 1994) notes that the *proviso* of subjecting these obligations to national legislation is unusual. The objectives of the article could be defeated since the wording implies that existing national legislation will take precedence. It also could be taken to imply that these concerns of Indigenous Peoples can be respected and preserved without addressing outstanding issues of Indigenous Peoples' rights to land and biological resources. It is obvious that such communities cannot continue these traditional practices in isolation from the land and biological resources that they need (IUCN, 1994), and this recognition would be consistent with a growing body of international obligations such as *ILO 169* and the *Draft Universal Declaration on the Rights of Indigenous Peoples*.

If Article 8(j) is to be given legal meaning, which respects the concerns of Indigenous Peoples, some fundamental issues must be addressed:

- Indigenous concepts of conservation and sustainable use need to be much better understood by the wider national and international community. This should be facilitated by providing the resources, and access, for Indigenous Peoples to express these concepts directly in their own words. It also involves a recognition of wide cultural diversity even within small groups of Indigenous Peoples.
- Indigenous knowledge, innovation and practices are also poorly understood. Ethnographic and ethnobiological studies are limited and have not necessarily been undertaken for the policy purpose of conservation and sustainable use of biological diversity. The knowledge and practice is deeply embedded in Indigenous culture and appropriate research and policy development will need to be undertaken through Indigenous control or partnership.
- The expression "Indigenous and local communities embodying traditional lifestyles" needs to be critically considered. Many Indigenous Peoples, with strong traditional links and involvement, may be excluded from Article 8(j) because of their employment or where they live. In particular, it fails to consider the realities of contemporary Indigenous culture. The Australian Government briefing for the first meeting of the Conference of Parties stated that Australia considers that the phrase "embodying traditional lifestyles" should not imply that such communities do not change and would include Indigenous Peoples who follow traditional customs but do so in a "non-traditional" way.
- Contracting parties are meant to promote wider application of Indigenous knowledge with the approach and involvement of relevant Indigenous people. The holder/s of the knowledge or technology may be an individual, group or community. This will make the participatory provisions difficult to implement without sound applied anthropological studies and co-operation from individuals and communities. It is unclear who can, or should, determine the issue of who are the "holders" of knowledge and technology.
- The provision for the "equitable sharing" of benefits, from the wider applications, raises the same issues discussed above. Will "equity" be determined from an Indigenous perspective and does it imply the recognition of cultural and intellectual property rights held by Indigenous Peoples? At the very least Indigenous Peoples will expect that the wider application of their knowledge, practices and technology would be preceded by recognition of Indigenous concerns in the first part of Article 8(j). The analogy is with the idea of a "trust." If national governments are to use Indigenous knowledge, innovations and practices, for the wider public "good," then there should be a clear obligation towards the Indigenous Peoples who have developed them. It would be against the intent (see Preamble)

of the Convention to construe Article 8(j) purely as a means of appropriating Indigenous knowledge without reciprocity. The legal and practical forms of this reciprocity remain to be worked out under the Convention.

Article 10(c) requires contracting parties to "Protect and encourage customary use of biological resources in accordance with traditional cultural practices that are compatible with conservation and sustainable use requirements."

Most of the Articles in the Convention recognise that non-Indigenous laws, policies and practices will change as we learn more about biodiversity and strategies to protect it. Indigenous culture has always been subject to some change. Indeed, this is why some of their biodiversity strategies and protective systems have been so effective. If expressions such as "customary use" and "traditional cultural practices" are interpreted as protecting only past, or existing, uses and practices this would deny contemporary Indigenous self determination and undermine many of the purposes of the Convention. The relevant focus is Indigenous sustainable use. Judgments about "traditionality" will impede Indigenous co-operation on these issues.

Some of these specific Indigenous Peoples' issues have been dealt with in a Work Programme of the CBD Conference of Parties (COP) from 1996 onwards. The COP (VI/10) requested that the CBD Working Group on Article 8(j) address the issue of sui generis systems for the protection of traditional knowledge and identified the following issues on which to focus:

- Clarification of relevant terminology;
- Compiling and assessing existing Indigenous, local, national and regional *sui generis* systems;
- Making available this compilation and assessment through the clearing-house mechanism of the Convention;
- Studying existing systems for handling and managing innovations at the local level and their relation to existing national and international systems of intellectual property rights, with a view to ensure their complementarity;
- Assessing the need for further work on such systems at the local, national, regional and international levels;
- Identifying the main elements to be taken into consideration in the development of *sui generis* systems; and
- The equitable sharing of benefits arising from the utilization of traditional knowledge, innovations and practices of Indigenous and local communities, taking into account the work carried out by the Intergovernmental Committee on Intellectual Property and Genetic Resources, Traditional Knowledge and Folklore with a view to promote mutual supportiveness, and existing regional, subregional, national and local initiatives (Working Group on Article 8(j), et al., UNEP/CBD/COP/6/20, para. 34).

The latest COP and Ad-Hoc meetings (2003) focused on improving both conceptual approaches and regulatory and implementation strategies for better State application of Article 8(j).[3] Recognition of the following illustrate the widening of international legal concepts: definitional problems of Article 8(j), inappropriateness of codifying traditional knowledge by western scientific methods and also intellectual property law outside a *sui generis* approach, and the need for research ethics to formulate codes of conduct when studying traditional knowledge and customs.

Recommendations for practical improvements include: strengthening importance of participation by Indigenous communities, capacity building within States with the aid of indicators to measure retention and loss of traditional knowledge and cultures, requesting States to identify key threatening processes against the intentions of this Article, and adherence to structured report writing to comprehensively illustrate the nations position on the application of Article 8(j).

Article 6 requires Contracting Parties to develop national strategies, plans or programmes for the conservation and sustainable use of biodiversity. This is one of the most important obligations for implementation, under the Convention.

The identification and monitoring provisions (Article 7) involve national surveys or inventories of biological diversity. Article 14 deals with environmental impact assessment and minimising adverse impacts. Those activities must involve affected Indigenous Peoples in a significant way. These Articles will provide important bases for Indigenous participation in planning for biodiversity conservation at the national and local levels and complement the rights in Articles 8(j) and 10(c) of the CBD.

CBD Intellectual and Cultural Property Issues

A close analysis of the CBD reveals a serious risk that Indigenous Peoples will be seen as a "resource" for biological diversity rather than as Peoples who hold legal and cultural rights in relation to it. This poses serious ethical and practical issues in seeking to involve Indigenous Peoples. Indigenous knowledge is a considerable source of wealth in the pharmaceutical and other industries and a constant focus of research activity. At the present time, virtually none of the profits are returned to Indigenous Peoples. Several Articles in the CBD are primarily concerned with promoting commercial access to genetic resources and promoting the commercial access and transfer of technology. The relevant articles (15 and 16) make no specific provisions for Indigenous Peoples and they have to be read in the context of the earlier articles (such as 8(j)).

The promotion of access to genetic resources and proposals to patent genes, could eventually deny Indigenous people the biological resources they have managed for thousands of years. These issues are of concern to Indigenous and non-Indigenous Peoples. The provisions for increased access to genetic resources need to be considered in the context of proposals to patent animal and life forms with the possible reduction in biodiversity because of monopolies in ownership and control.

The provisions for the access and transfer of knowledge and technology (Articles 16, 17 and 18) include Indigenous and traditional knowledge and technology. The term "technology" can encompass such knowledge and technology (in Article 16) and it is explicitly referred to in Articles 17 and 18. The only basis for Indigenous "control", "participation" and "benefit" is contained in Article 8(j). The scene is set for wide use of Indigenous knowledge and practices relating to biodiversity.

However, few jurisdictions have developed legislation and Codes of Conduct which will ensure that some of the benefits are returned to Indigenous communities.

One strategy for Indigenous Peoples is to use existing intellectual property law for their own benefit where possible. At the same time, they are pushing for changes to intellectual property laws to encompass their cultural concerns and collective forms of "ownership". Proposals for national biodiversity laws will need to consider existing and alternative frameworks for addressing the intellectual and cultural property rights appropriate for Indigenous Peoples.

The language of the CBD, however, encourages rather than obliges States to protect the rights of Indigenous Peoples and to develop national legislation to respect, preserve and maintain the knowledge, innovations and practices of traditional Peoples. National laws are required to protect the intellectual property and traditional resource rights of communities embodying traditional lifestyles. The active approval by and involvement of Indigenous, traditional and local communities in conservation and development activities is fundamental for the successful implementation of the CBD. Furthermore, equitable sharing of benefits and development of effective forms for compensation are critical, as are mechanisms to protect local communities from the adverse effects of external technologies that weaken their effectiveness in conservation efforts.

Much research is needed to understand the effectiveness of traditional technologies. Research, monitoring, and inventory criteria, priorities, and methods need to be guided and controlled by Indigenous and local communities. Finally, to successfully implement the provisions of the CBD, financial mechanisms will have to be made available to Indigenous and local communities.

INTERNATIONAL CONVENTION TO COMBAT DESERTIFICATION IN COUNTRIES EXPERIENCING SERIOUS DROUGHT AND/OR DESERTIFICATION

The *International Convention to Combat Desertification in Countries Experiencing Serious Drought and/or Desertification* 1994, does not specifically mention Indigenous or traditional Peoples, it does give significant emphasis to local Peoples, local communities, local populations, and local organisations. It also recognises local and traditional knowledge, know-how, practice, and skills. These are mentioned throughout the text, but given prominence in Article 17.

Principles and Articles

Article 9 calls for the preparation and implementation of National Action Programmes (NAPs), including regional and subregional programs, to combat desertification. These programs call for a participatory process based on lessons from field action, as well as the results of research. These NAPs are similar to the national plans called for in the CBD, and they should specify the respective roles of government, local communities and land users and the resources available and needed.

Article 10(3) permits NAPs to include such things as local and regional early warning systems, strengthening of food security systems (including storage and marketing), and establishment of "alternative livelihood projects that could provide incomes in drought prone areas". Article 16 deals with information collection, analysis and exchange. Data banks and networks are called for that address the needs of local communities and ensure the adequate protection and return for local and traditional knowledge that is provided.

Article 17 deals with research and development and requires Signatories to:

> protect, integrate, enhance and validate traditional and local knowledge, know-how and practices, ensuring, subject to their respective national legislation and/or policies, that the owners of that knowledge will directly benefit on an equitable basis and on mutually agreed terms from any commercial utilisation of it or from any technological development derived...

Article 18 treats the subjects of transfer, adaptation and development of technology. Article 19 calls for capacity building, educational, and public awareness measures, including training in alternative technologies, conservation, and research. The Article also calls for action...

> by fostering the use and dissemination of the knowledge, know-how and practices of local people in technical cooperation programmes, wherever possible; by adapting, where necessary, relevant environmentally sound technology and traditional methods of agriculture and pastoralism to modern socio-economic conditions...

This raises serious issues about what (if any) mechanisms will be used to protect Indigenous rights to intellectual and cultural property.

Articles 20 and 21 deal with financial resources and mechanisms and call for assistance from the Global Environmental Facility (GEF) and international lending institutions. They also recognise debt swaps "and other creative mechanisms." Article 21(3) calls on States to utilise participatory processes involving non-governmental organisations, local groups and the private sector, in raising funds, in elaborating as well as implementing programmes and in assuring access to funding by groups at the local level.

A "Global Mechanism" is also established in Article 21(5) to guide, inform, advise and facilitate States in their attempts to finance relevant programs and projects. Such a mechanism should address the specific concerns of Indigenous, traditional, and local communities.

In general, the *Convention to Combat Desertification* (COD) offers significant promise to some local communities and, in as much as Indigenous communities are local communities, this should not go unnoticed. As with the CBD, the call for wider use and application of local and traditional know-how, knowledge, practice, and skills may be well-intentioned, but without adequate provisions for protection and just compensation for such use, such a call is dangerous.

The provisions of the *Convention to Combat Desertification* go further than the CBD in their protection of intellectual, cultural, and scientific property rights. However, no specific mechanisms are proposed, nor even any process to develop such provisions. This situation should signal to Indigenous and local communities an even greater and more urgent need to develop their own guidelines for technology access, transfer, and protection.

Although the COD has somewhat stronger wording in support of local communities and their traditional knowledge, know-how, and practices, it suffers the same problems as the CBD. These include extension of State sovereignty over traditional resources. It recognises and only tacitly guarantees Indigenous rights. Furthermore, the COD does not enjoy the general application and international grassroots support of the CBD. Consequently it is less likely that the COD will be moulded by public sentiment in favour of Indigenous Peoples. Nonetheless, there is no reason that such pressures could not be mobilised, since the *Convention to Combat Desertification* is yet another major international agreement that can and should serve as a pivotal point for the integration of human and environmental rights.

NATIONAL RECOGNITION OF CUSTOMARY LAW

Some nations have recognised the customary laws of Indigenous Peoples to use and manage their knowledge and biological resources through the common law or through statute. In the Canadian (*Calder* v. *Attorney General for British Columbia*, 1973), New Zealand (*Te Weehi* v. *Regional Fisheries Officer*, 1986) and Australian experience, the recognition of native title has encompassed related customary laws for the use of land and natural resources. In *Mabo v Queensland* (*Mabo* v. *State of Queensland*, 1992), the Australian High Court recognised the existence of native title which entitles Indigenous Peoples in Australia to the use and enjoyment of ancestral lands in accordance with their unique laws and customs. Four judges—Brennan, Dean, Gaudron, and Toohey, JJ—explicitly rejected a narrow view of "traditional law or custom." Justice Brennan stated:

> Of course in time the laws and customs of any people will change and the rights and interests of the members of the people among themselves will change too. But so long as the people remain as an identifiable community, the members of whom are identified by one another as members of that community living under its laws and customs, the communal native title survives to be enjoyed by the members according to the rights and interests to which they are respectively entitled under the traditionally based laws and customs, as currently acknowledged and observed...It is immaterial that the laws and customs have undergone some change since the Crown acquired sovereignty provided the general nature of the connection between the Indigenous people and the land remains.

A Canadian Supreme Court case has decided that the Canadian Federal Government owes a fiduciary obligation to Indigenous Peoples when they dispose of ancestral or reserve land (*Guerin* v. *the Queen*, 1983). This Court also held that native title includes practices that form an integral "part" of an Indigenous community's distinctive culture such as the Indigenous use of fisheries. An argument can be made that native title necessarily includes Indigenous management of marine resources, wildlife, natural resources, land and waters, together with the associated intellectual and cultural property.

In the context of Canada, the recognition of native title has led to the negotiation of some modern treaties which settle Indigenous land claims, allocate rights to natural resources (and royalties from their exploitation) and set up comprehensive regimes for Indigenous participation in environmental assessment, development decisions and the management of land, seas, natural resources and wildlife (Canadian Regional Agreements) (Netheim et al., 2002). Native title rights and Regional Agreements are granted constitutional protection (S. 35(i) of the Canadian Constitution Act, 1867 as amended in 1982).

The jurisprudence whereby longstanding treaties between governments and Indigenous Peoples are being interpreted in the modern context also gives greater emphasis and recognition to Indigenous customary laws relating to the sustainable use and management of biological resources and associated intellectual and cultural property.

CONCLUSION

The international law and policy relating to human rights, including the intellectual and cultural property rights, of Indigenous Peoples, is evolving rapidly. It currently falls short of the *sui generis* system for protecting the integrated rights of Indigenous people that Posey advocates. However, it is becoming increasingly clear that this type of system accords closely with a growing body of international law and policy specifically relating to Indigenous rights and the aspirations of Indigenous Peoples for self determination.

In many contexts the global recognition of the integrated human and environmental rights of Indigenous Peoples would facilitate SD and benefit the wider community. This reality should not obscure the many conflicts over priorities and strategies that can occur between the Indigenous Peoples and the environmental movement. Much work remains to be done in understanding human rights, environment and development rights and specific Indigenous rights. The sometimes strained relationship between the environmental and Indigenous rights movements raises fundamental ethical, legal and moral issues.

In some situations there is desperate urgency to meet basic needs and alleviate poverty in Indigenous communities. There are limited experiences with poverty reduction specifically directed to Indigenous Peoples. The International Working Group for Indigenous Affairs has published a series of articles discussing needs based approaches vs rights-based approaches (IWGIA, 2003) Maria Quispe concludes that this dichotomy can be a trap when planning economic and social policy involving Indigenous Peoples. A rights—based approach usually tries to make People's needs become a right; promote Peoples knowledge about the existence of their rights; and develop political strategies to make people's rights become reality (IWGIA, 2003)

The argument continues that we should never be afraid to re-think certain rights and how we try and make them mean something beneficial. Traditional rights and development approaches have failed Indigenous Peoples and they are increasingly irrelevant to Indigenous self-determination and SD:

> There is a need to further develop an integrated approach to working with Indigenous Peoples' rights and development simultaneously as it would contribute to the on-going work on designing effective poverty reduction strategies (IWGIA, 2003).

KEY

Legally Binding Agreements in Force (with number of State Parties)

CBD:	Convention on Biological Diversity (1992)
CDW:	Convention on the Elimination of all Forms of Discrimination Against Women
CERD:	Convention on the Elimination of all Forms of Racial Discrimination (1966)
CG:	Convention on the Prevention and Punishment of the Crime of Genocide (1948)
CRC:	Convention on the Rights of the Child
GATT:	Final Act Embodying the Results of the Uruguay Round of Multilateral Trade Negotiations (1994)
ICESCR:	UN International Covenant on Economic, Social and Cultural Rights (1966)
ICCPR:	UN International Covenant on Civil and Political Rights (1966)
ILO169:	International Labour Organisation Convention 169: Convention Concerning Indigenous and Tribal Peoples in Independent Countries (1989)

NLS: NATIONAL LAWS

RC:	Rome Convention for the Protection of Performers, Producers of Phonograms and Broadcasting Organisations (1961)
UNESCO:	UNESCO Convention Concerning the Protection of the World Cultural and Natural Heritage (1972) (WHC)
UNESCO:	UNESCO Convention on the Means of Prohibiting and Preventing the Illicit Import, Export and Transfer of Ownership of Cultural Property (1970) (CCP)

UPOV: International Union for the Protection of New Varieties of Plants (1961, revised in 1972, 1978 and 1991)
WIPO: The World Intellectual Property Organisation, which administers international IPR agreements, such as:
The Convention of Paris tor the Protection of Industrial Property (1883, revised most recently in 1967)
The Berne Convention for the Protection of Literary and Artistic Works (1886)
The Madrid Agreement Concerning the International Registration of Trademarks (1891)
The Lisbon Agreement for the Protection of Appellations of Origin and their International Registration (1958)
The Patent Cooperation Treaty (1970)

Non-legal Agreements

DDHRE: UN Draft Declaration of Principles on Human Rights and the Environment (1994)
DDRIP: UN Draft Declaration on the Rights of Indigenous Peoples (formally adopted by the UN Working Group on Indigenous Populations in July 1994)
DHRD: UN Declaration on the Human Right to Development (1986)
FAO-IUPGR: FAO International Undertaking on Plant Genetic Resources (1987 version)
RD: Rio Declaration (1992)
UDHR: Universal Declaration of Human Rights (1948)
UNESCO-F: UNESCO Recommendations on the Safeguarding of Traditional Culture and Folklore (1989)
UNESCO-PICC: UNESCO Declaration on the Principles of International Cultural Cooperation (1966)
UNESCO-WIPO: UNESCO-WIPO Model Provisions for National Laws on Protection of Expressions of Folklore Against Illicit Exploitation and Other Prejudicial Actions (1985)
VDPA: UN Vienna Declaration and Programme of Action (1993)

NOTES

[1] Under the Sub-Commission on the Protection and Promotion of Human Rights, the Working Group on Indigenous Populations comprised of a panel of experts created a Draft Declaration of Indigenous Rights and provided a platform for Indigenous issues. Under the direction of the Commission on Human Rights, the Working Group on the Draft Declaration on the Rights of Indigenous Peoples continues the shaping of the Draft Declaration for adoption of the Commission and is comprised of member states with the consultation and extensive participation of Indigenous groups.

[2] See http://www.unesco.org/whc/nwhc/pages/doc/main.htm for more information about this and all other World Heritage sites.

[3] See also Ad-Hoc Open Ended Inter-Sessional Working Group on 8(j) Provisions of the Convention on Biological Diversity, "Composite Report on the Status and Trends Regarding the Knowledge, Innovations and Practices of Indigenous and Local Communities Relevant to the Consideration and Sustainable User of Biodiversity" UNEP/CBD/WG8J/3/4, 28th Sept 2003, see Website <http://www.biodiv.org/doc/meeting.asp?wg=WG8J-03

REFERENCES

Aboriginal and Torres Strait Islander Commission Indigenous Reference Group (1998) Our culture: our future, written and researched by Janke T, Frankel M, Michael Frankel & Company and Terri Janke, pp 47–48

Agreement on Trade Related Aspects of Intellectual Property Rights (1994, April 15) Marrakesh Agreement Establishing the World Trade Organisation, Annex 1C, GATT Doc. MTN/FAII- A1C, 33 I.L.M. 1197

Brownlie I (1992) The rights of peoples in modern international law. In: Crawford J (ed) The rights of peoples. Clarendon Press, Oxford

Clinton RN (1990) The rights of indigenous peoples as collective group rights. Ariz L Rev 32(4):739–747

Cocoyoc Declaration, 1974—adopted by the participants in the UNEP/UNCTAD symposium on "Patterns of Resource Use, Environmental and Development Strategies", Cocoyoc, Mexico, October 8–12, 1974.

Craig D, Ponce DN (1995) Indigenous peoples' rights and environmental law. In: Lin S, Kurukulasuriya Y (eds) UNEP's new way forward: environemental law & sustainable development. UNEP, Nairobi

Craig D, Robinson N, Koh KL (eds) (2002) Capacity building for environmental law in the Asian and Pacific region: approaches and resources (Two Volumes), Asian Development Bank, Manilla

Daes E (1993) Study on the protection of cultural and intellectual property of indigenous peoples. Report of a Special Rapporteur to the UN Sub-Commission on Prevention of Discrimination and Protection of Minorities, ECOSOC, E/CN.4/Sub.2/1993/28:31–32

Declaration of San Jose UNESCO, meeting of experts on ethnodevelopment and ethnocide in Latin America, San Jose, Costa-Rica, 7–11 December 1981.

Food and Agriculture Organisation of the United Nations (FAO) (1989) Resolution 4/89. Twenty-fifth Session of FAO Conference, Rome

Hansen SA, VanFleet J (2003) Traditional knowledge and intellectual property: a handbook on issues and options for traditional knowledge holders in protecting their intellectual property and maintaining biological diversity. AAAS, Washington

IUCN Environmental Law Centre (1994) The convention on biological diversity: an explanatory guide. IUCN Biodiversity Program, Bonn

IUCN Inter-Commission Task Force on Indigenous Peoples (1997) Indigenous peoples and sustainability cases and actions, IUCN, The Netherlands

IUCN Draft Covenant on Environment and Development, (2nd edn) Commission on environmental law. (2000) Available online: http://www.iucn.org/themes/law

International Work Group on Indigenous Affairs (IWGIA) (2003) Indigenous Poverty: an Issue of Rights and Needs. Indigenous Affairs 1.

Johnson M (1992) Lore: capturing traditional environmental knowledge. Dene Cultural Institute/IDRC

Meeting of Experts on Human Rights and Environment (1994) Geneva, Ottawa 16–18 May

Netheim G, Craig D, Meyers D (2002) Indigenous governance structures: a comparative analysis of land and resource management rights. Australian Institutions for Aboriginal and Torres Strait Islander Studies. Canberra.

Posey D (1996) Traditional resource rights: international protection and compensation for indigenous peoples and local communities. IUCN, Gland, Switzerland

Posey DA (1999) Introduction: culture and nature – The inextricable link. Cultures and spiritual values of biodiversity, UNEP

Rural Advancement Foundational International (1994) Conserving indigenous knowledge: integrating two systems of innovation. Commissioned by the United Nations Development Programme, New York

Shutkin WA (1991) International human rights law and the eEarth: the protection of indigenous peoples and the environment. Virginia J Int'l L 31(3):479–511

The [Canadian] Constitution Act, 1867 as amended in 1982, Schedule B (Canadian Charter of Rights and Freedoms) - Section 35(1)

The World Conservation Union (2001) World wide fund for nature and united nations environment programme. Caring for the Earth

UN (1992) Conference on Environment and Development: Convention on Biological Diversity, 5 June 1992 UNEP/Bioprospecting.Div./N7BINC5/4, reprinted in 31 I.L.M. 818 (entered into force 29 Dec 1993)

UN ESOSOC (1986) Study of the Problem of Discrimination Against Indigenous Populations, United Nations Economic and Social Council, E/CN.4Sub.2

United Nation (2002) Meeting of Experts on Human Rights and the Environment, OHCHR

UN General Assembly Resolution 128, UN GAOR (41st mtg), UN Doc. A/Res/41/128 (1986)

UNEP (1992) Saving our planet: challenges and hopes, Nairobi, Kenya

UNESCO (2003) General Conference 32nd Session, Press Release. Available online:http://portal.unesco.org/en/ev.php@URL_ID=16783&URL_DO=DO_TOPIC&URL_SECTION=201.html

United Nations (1972) Conference on the human environment. Stockholm

Ward E (1993) Indigenous peoples between human rights and environmental protection. Danish Human Rights Centre, Copenhagen

Working Group on Article 8(j) and Related Provisions of the Convention on Biological Diversity, "Conference of the Parties", UNEP/CBD/COP/6/20, Paragraph 34

World Summit on Sustainable Development (2002) Plan of Implementation

Yamin F, Posey D (1993) Indigenous peoples, Biotechnology and intellectual property rights. Rev Eur Commun & Int'l Envtl L 2(2)

CASES

Guerin v. the Queen (1983) 144 DLR (3rd) 193 S.C.
Mabo v. State of Queensland (1992) 175 CLR 1
Oposa vs Factorian (1993) GR No 101083, 30th July 1993, 224 SCRA 792
R v. Sparrow (1990) 1 SCR 1, 75
Te Weehi v. Regional Fisheries Officer (1986) 1 NZLR 680

IKECHI MGBEOJI

CHAPTER 6

LOST IN TRANSLATION? THE RHETORIC OF PROTECTING INDIGENOUS PEOPLES' KNOWLEDGE IN INTERNATIONAL LAW AND THE OMNIPRESENT REALITY OF BIOPIRACY

INTRODUCTION

It is no longer debatable that human intervention by indigenous[1] and traditional peoples all over the world played and continues to play a crucial role in the development and sustenance of plant species and genetic resources across the globe (Halewood, 1999; Four Directions Council, 1996; Kloppenburg, 1988). In spite of the enormous contributions of indigenous and traditional peoples to plant species diversity, however, there remains a huge South[2]–North[3] divide in the apportionment or enjoyment of not only the material benefits of such resources, but more importantly, in the way law acknowledges and protects the intellectual contributions that produced or improved the genetic resources in question. In this chapter, I argue that the optimistic rhetoric in some international law instruments purporting to protect indigenous peoples knowledge has not been matched with action. The unfortunate reality, I argue, is that the intellectual contributions of traditional and indigenous peoples to plant species development and sustenance have largely remained unrecognised, unrewarded and ruthlessly exploited by both the general framework of international law and various domestic intellectual property law institutions in the industrialized countries of the North.

Rather than being vigorously protected and genuinely respected, traditional and indigenous peoples' knowledge systems have been ridiculed by the dominant epistemology and legal culture. To add insult to injury, the products of indigenous and traditional knowledge frameworks and cultures are continually appropriated through legal constructs and commercial rules fashioned by the industrialized North. Consequently, the unauthorized appropriation of indigenous peoples bio-cultural knowledge and resources, otherwise known as "biopiracy"(Mgbeoji, 2005; Dutfield, 2004), has proceeded and flourished as if it were a lawful and morally justifiable enterprise (RAFI, 1995a, 1999; Shiva, 1999; Villaema, 2000).

In spite of the egregious historical injustices done to indigenous and traditional peoples in both the colonial age and the extant economic inequities in the global legal and economic order, the phenomenon of biopiracy has not been fully appreciated as part of a continuous and persistent egregious assault on the culture, worldview, and sustenance of indigenous peoples across the globe (Tejera, 1999). For scholars and policy-makers to fully apprehend the true nature of biopiracy, I argue in this paper that contemporary intellectual property law must be deconstructed and understood as a congealed manifestation of European culture and values (Sarma, 1999; RAFI, 1995a, 1995b; Orr, 1999; Osava, 1999). As a social institution, law often, but not necessarily always, reflects the biases and preferences of the dominant segments of society (Hunn, 1993). Given the historical marginalization and subjugation of indigenous peoples, it goes without saying that contemporary international intellectual property regimes have ignored indigenous peoples values and interests.

Interestingly and for good historical and modern reasons, it is the general opinion of indigenous and traditional peoples that given the Eurocentric biases and preferences of international law (Mutua, 2001; Anghie, 2004), prevailing models of intellectual property regimes, especially patents and copyrights, have wrought disaster on indigenous peoples, their cultures and the products of their epistemic frameworks. As the Bellagio Declaration on the Rights of Indigenous Peoples notes:

> [I]ntellectual property laws have profound effects on issues as disparate as scientific and artistic progress, biodiversity, access to information, and the cultures of indigenous and tribal peoples. Yet all too often those laws are constructed without taking such effects into account, constructed around a paradigm that is selectively blind to the scientific and artistic contributions of many of the world's cultures and constructed in fora where those who will be most directly affected have no representation.

When modern intellectual property regimes were created and institutionalized, the consent and opinion of indigenous and traditional peoples across the world was neither sought nor obtained. Similarly, when the dominant intellectual property regimes was imposed on colonized peoples and territories, neither the consent nor permission of indigenous and traditional peoples was sought (Mgbeoji, 2003). Since I have already dealt with some of these issues elsewhere (Mgbeoji,2001a), this chapter moves beyond the issue of appropriation of indigenous and traditional peoples' bio-cultural knowledge to examine the present status of the colonial neglect and denigration of traditional and indigenous peoples' knowledge systems. It also traces the struggles undertaken by indigenous peoples and their sympathizers to emancipate indigenous and traditional peoples' knowledge from the shackles of an arrogant and homogenizing Eurocentric international intellectual property regime.

This chapter also argues that the relegation of indigenous and traditional peoples' knowledge to the realms of magic and irrationality in the colonial era was a function of the racist and prejudiced biases of the dominant legal and cultural order. This unfortunate era of global history, again, affords ample proof of international law's predilection to enact as law and proclaim as normative, the biases and preferences of the North at the expense and exclusion of peoples of the global South. More importantly, I argue that in spite of the well meaning attempts by some academic institutions, research organizations and non-governmental organizations, *et cetera*, to draw up ethical guidelines on access to and use of indigenous and traditional peoples' knowledge, international law remains embarrassingly weak and ineffectual on the protection of indigenous and traditional peoples knowledge.

While modern international intellectual property regime robustly protects Western forms of intellectual property, the status of indigenous and traditional peoples' knowledge within the general framework for the protection of intellectual property hovers somewhere between the moribund and the comatose. In many respects, indigenous and traditional peoples' knowledge remains easy picking for free and ruthless exploitation by anyone or institutions possessed of the requisite cunning and infrastructure.

In explaining the central thrust of this chapter, three clusters of issues are dealt with. The first cluster of issues evaluates the normative impact of the inattention by scholars to the radical impact that plants and traditional knowledge on the uses of plants have had on shaping and reshaping global economic and legal order. The second phase lays bare, albeit briefly, the influence of racial prejudice on the construction of protectable and respectable forms of intellectual property. Third given that traditional and indigenous peoples' knowledge was often dismissed as unscientific, such knowledge was freely appropriated.

The erroneous notion that traditional and indigenous peoples' knowledge was a free good in the public domain stemmed from the colonialist and racist idea that indigenous and traditional peoples were intellectually incapable of making innovations worth protecting through the mechanisms of the dominant intellectual property regimes. The persistent unwillingness of the major intellectual property regimes to accord intellectual property right protection or recognition to innovations or products arising from traditional and indigenous knowledge systems was thus a function of both philosophical incompatibility and attitudes of epistemological inferiority.

However, in recent times, this attitude, as reflected in both international and domestic law, has witnessed a slight change. Modern international law instruments such as the Convention on Biological Diversity (CBD) now recognize the empirical and scientific character of indigenous peoples' knowledge. Similarly, some states have begun to enact domestic laws protecting traditional and indigenous peoples' knowledge. Arguably, it is becoming increasingly fashionable for state-parties to the CBD and similar conventions to fashion domestic legal regimes for the protection of indigenous peoples' knowledge and for the equitable sharing of the benefits of such knowledge. The progress, however, has been piece-meal, ineffectual and devoid of the same rigour with which Western intellectual property rights are compulsorily enforced across the globe.

This chapter is divided into three substantive parts. Part 1 highlights the inattention by scholars to the contributions of indigenous and traditional peoples, particularly in respect of plant species development. Part 2, argues that scholarly neglect of the intellectual and scientific merits of traditional and indigenous peoples' knowledge may be largely socio-cultural in nature. This neglect was probably a function of antiquated cultural, gendered, and racist denial of the intellectual input of traditional farmers and breeders, particularly women in the improvement of plants. In other words, Western scholarly inattention to the attainments of traditional knowledge systems was arguably, a direct product of colonial and pre-modern attitudes that traditional and non-European peoples were inferior in their intellectual capacity and could not have wrought any rational improvements in plant species use and development. Part 2 further deals with the gendered undercurrents which shaped the institution or paradigm of Western science and banished traditional or indigenous peoples' knowledge systems to the realms of magic and "old wives tales." This part also evaluates the historical and cultural forces that placed Western empiricism on the pedestal while dismissing non-Western empiricism as irrational and unscientific.

In sum, Parts 1 and 2 seek to demonstrate how the various mechanisms of appropriation of indigenous and traditional peoples' knowledge (biopiracy) derive their vitality and juridical support from the twin-processes of socio-cultural denigration and non-justiciability of claims of indigenous and traditional peoples in claims pertaining to their intellectual property. Under this socio-legal construct, resources and products derived from traditional and indigenous knowledge frameworks were considered a global good, belonging to no one. The erroneous notion that such knowledge was free goods afforded an untenable basis for the appropriation of traditional and indigenous peoples' knowledge (Stenson and Gray, 1999; cf. Onwuekwe, 2004).

In this context, Part 3 examines the normative weight and rigour of modern thrusts of international and domestic laws that seek to validate the claims of indigenous peoples' knowledge systems for respect and legal protection. The chapter thus concludes with the observation that although the CBD and other modern developments in international and domestic law have substantially dismantled the appropriative functions of Western epistemology and supporting intellectual property regimes, more work needs to be done not only by individual states but within the framework of international law if the intellectual property of indigenous and traditional peoples is to be protected and accorded respect.

INDIGENOUS PEOPLES, COLONIALISM AND THE ROLE OF PLANTS IN THE GLOBAL ORDER

In this age when concerns about the egregious maltreatment and in many cases, destruction, of indigenous peoples and erasure of their contributions have assumed prominence in global discourse (Wiessner, 1999), it is easy to forget that some decades ago, indigenous peoples occupied the outer peripheries of issues addressed by powerful members of the international community. With the formal end of colonialism and the emergence of what some scholars have termed the "post-colonial" (Fitzpatrick and Darian-Smith, 1999) era, indigenous and traditional peoples have increasingly and consistently shared intellectual and political space with other members of the global community. This shared space has resulted in some landmark court decisions recognizing aboriginal titles to land in Canada, Australia, New Zealand, and several other colonial outposts. While it is permissible to celebrate these "victories", it must be borne in mind that indigenous and traditional peoples are still shackled by colonizing forces and oppressive legal frameworks imposed on them since the age of conquest and colonization.

The Concept of Indigenous Peoples' Knowledge

Perhaps the most significant aspect of the renaissance of indigenous and traditional peoples' rights is the emergence of indigenous and traditional epistemic frameworks as valid and respectable approaches to the acquisition, transfer, development and retention of knowledge. Before examining the contemporary status of indigenous peoples' knowledge, it is pertinent to clarify the concept of indigenous peoples' knowledge itself. It is difficult to set boundaries between various forms of knowledge and various scholars are strongly of the view that there is no satisfactory definition of indigenous knowledge (Battiste and Henderson, 2000). Although the concept of indigenous knowledge is often mentioned in relation to knowledge centered on the ecological worldviews of colonized peoples and societies, various epistemological frameworks have long engaged in inter-penetrations and exchanges of worldviews and narrative frameworks (Semali and Kincheloe, 1999). The high degree of cultural cross-fertilization complicates theorizing on the nature of indigenous knowledge. As a result of the cultural interaction between various forms of epistemology, the distinction drawn in this chapter between Western epistemology and traditional and indigenous forms of knowledge is convenient, rather than purist.

This distinction does not imply the absence of "traditional and indigenous" knowledge in Western societies. Rather, the distinction drawn between Western epistemology and indigenous peoples' knowledge is a broad, generalized differentiation of the divergent conceptualizations of ways of seeking knowledge adopted and practiced by different cultures (Hunn, 1993). It is not a purist categorization but a convenient tool of analysis premised on anthropological studies, cultural differences, and power relations between the colonizing and the colonized.[4]

Since modern colonization, the mainstreaming of European norms and the "othering" of non-Western forms of civilization and worldviews banished the latter to the fringes of the forgotten. Consequently, the concept of indigenous peoples' knowledge is better understood within the context of colonialism and the disruption and truncation of the natural development of pre-colonial epistemological frameworks. In this paradigm the attempt to use Western empiricism as the measuring rod for indigenous knowledge may be construed as a continuation of the colonial entrapment and marginalization of colonized cultures and peoples. Hence, the concept of indigenous knowledge is part of the legal and socio-cultural claims of indigenous peoples to shared equality, dignity, and respect with other peoples across the world. As knowledge systems are in flux, it also follows that the distinction between indigenous knowledge and Western scientific knowledge would equally change with the times (Oguamanam, 2004).

The Neglect of Indigenous Peoples' Contributions to Plant Diversity

Although it is arguable that there is perhaps a trend towards the romanticization of indigenous peoples' knowledge (Johannes, 1989), it is significant that this development is a product of social ferment across various sites at which power and law are negotiated. This radical change in the power relations between indigenous knowledge systems and the dominant epistemic narrative has not gone unnoticed in scholarly circles (Agrawal, 1999). In spite of these modern radical swings, it bears noting that for a long time the merits of indigenous knowledge was hardly countenanced by scholars nor narrated by establishment historians in the North. Remarkably, analysts of colonial history largely tended to focus on the looting, pillaging and theft of artifacts of gold, silver, ivory, wood, *et cetera* by the colonialists. Even Karl Marx's revolutionary and profound insight into and analysis of material dialecticism in global capital history, paid little attention to the role of indigenous knowledge systems in shaping the colonial age and the enormous contributions to global food supply and plant genetic diversity made by indigenous and traditional peoples.[5]

As a consequence of this neglect or oversight, the critical roles played by plants developed by indigenous peoples in re-shaping global political and economic order has been under-appreciated . Yet, in the colonial enterprise, indigenous ecological knowledge and plants developed by local farmers in the so-called "primitive and backward" territories played and continue to play radical and determinative roles in food security, medicine, ecology, and several other human needs (Fowler and Mooney, 1990). There is hardly any doubt that indigenous and traditional knowledge systems made their greatest contributions to plant species diversity and agriculture. Therefore, in evaluating the nature of power relations between indigenous and traditional knowledge systems vis-à-vis the dominant Western epistemological framework, emphasis should be placed on the legal regime on plants. In this context, it may be argued that the reason for the remarkable scholarly inattention to the role of plants in shaping the global order is that unlike gold and silver, plants are renewable resources. It is also possible that another reason is the near prosaic nature in which plants were transferred from one territory to the other. Save in exceptional circumstances, the history of plant transfer is rather dull and bears little similarity to the high drama, chicanery, and heroics usually associated with state-sponsored looting and pillaging of the pre-colonial Third World by swashbuckling European explorers, pirates, and *conquistadors* (Mgbeoji, 2003).

Simply put, plant theft seems benign, if not mundane. It seems unlikely that if the famed state-sponsored pirate, Sir Francis Drake had upon return from his voyages presented the British Queen with a bowl of cotton seeds, rubber-seeds or peanuts and the like, that a knighthood would have been conferred on him. Yet, economic plants, most of which were developed by indigenous and traditional peoples in pre-colonial times, have always formed the substratum on which diverse human civilizations have prospered and rested. For example, since the times of Christopher Columbus:

> [T]he New World supplied new plants of enormous culinary, medicinal, and industrial significance: cocoa, quinine, tobacco, sisal and rubber. More than this, the Americas also provided a new arena for the production of Old World's plant commodities (e.g., spices, bananas, tea, coffee, sugar, indigo) (Kloppenburg, 1988).

The profound implication here is that the asymmetrical movement of plant life forms from the areas occupied by indigenous and traditional peoples to colonial outposts redefined the structure and configuration of modern global economy, human population distribution, culture, science, and international legal order. For example, having destroyed its own forests (before the age of iron ships) the imperial might of the United Kingdom, largely dependent upon its command of the high seas, would have been otherwise were there no Indian forests to keep the British navy and merchant marine afloat. Similarly, it is remarkable that "a single coffee tree reaching the Amsterdam botanic gardens in 1706 from Ethiopia via Ceylon and Java became the basis for the New World coffee industry" (Lesser, 1997). Today, the culture of coffee consumption is so commonplace in Europe and the Americas that there is hardly any street in Europe or the Americas without a coffee shop.

Furthermore, and perhaps, more poignantly, the trajectory of modern American economy, jurisprudence and culture is primarily a function of the manner in which the young United States of America utilized or sought to exploit plant resources derived from indigenous and traditional peoples. The sustained and institutionalized use of enslaved labour rested on the cultivation of "exotic" plants like cotton, sugarcane, *et cetera*. Ultimately, the theory of racial inferiority of peoples of colour and the consequential contribution of that practice to the discourse and modern law on human rights would probably not have occurred if there were no sugarcane plantations, cotton-farms, *et cetera*. Neither the American Civil War nor the sea change in the jurisprudence of equality of humanity would have occurred or been developed without the contributions of plants to the slave system in the United States.

Plants and their associated uses, even when appropriated, have played radical roles in reconfiguring and influencing the ramifications of humanity and in re-shaping global politics, economics and international law. The appropriation of indigenous peoples' knowledge of the uses of plants fundamentally changed the world. The case of quinine is probably a perfect illustration of this phenomenon.[6] It is hardly debatable that it was quinine that enabled European colonizers to penetrate, survive and ultimately colonize the malaria-infested parts of Africa, Asia and Latin America. In effect, indigenous peoples' knowledge of the uses of plants have often played critical roles in redefining, reconfiguring and altering the global balance of power and indeed, affording the locus for the interrogation and re-examination of the inner core of legality and justice in the historical age.

Perhaps more perplexing than the inattention to the radical role of indigenous peoples' knowledge of the uses of plants in charting human civilization is the paucity of scholarly celebration of the enormous contributions of indigenous and traditional peoples to plant genetic diversity and conservation. This phenomenon may be traced to the centuries of denial by the dominant intellectual and scientific paradigms of the mental and intellectual capacities of non-Western cultures and societies. This attitude facilitated and condoned the appropriation of indigenous peoples knowledge of the uses of plants. Even where indigenous peoples' contributions to plant species diversity have been celebrated in the postcolonial age, it would seem that a significant part of the narration has been done by non-indigenous or traditional scholars; thus creating the impression that indigenous and traditional peoples lack the intellect, ability or capacity to tell their own stories (Battiste, 2000).

INTERNATIONAL LAW, INDIGENOUS PEOPLES AND THE CLASH OF EPISTEMOLOGIES

The colonization of non-Europeans was justified on the pillar of a theory of racial superiority of Europeans vis-à-vis the alleged inferiority of "the savages and primitives" of Africa, Asia, the Americas, Australia and New Zealand. The colonial enterprise was not only an economic and political mission, but also a legal concretisation of a racist agenda. Therefore, as Makau Wa Mutua (2000) has argued, as a predatory and racially motivated regime that validated the exploitation of non-European segments of humanity, the regime of international law "is illegitimate. It is a predatory system that legitimizes, reproduces, and sustains the plunder and subordination of the Third World by the West." It was on the basis of racial hierarchy of humanity that early international law justified the acquisition and colonization of large swathes of lands and cultures occupied by peoples which the colonizing Christians from Europe dismissed as "backward territories" (Lindley, 1969) and primitive peoples.

Although some Dominican clerics asserted the essential humanity of indigenous peoples of the Americas (Sanders, 1983), early international law, which was a product of Christian Europe, were fundamentally a congealed form of Christian bigotry and a congregation of the Christian brotherhood (Javis, 1991). In the words of Judge Mohammed Bedjaoui of the International Court of Justice, "this classical international law consisted of a set of rules with a geographical bias (it was a European law), a religious-ethical aspiration (it was a Christian law), an economic motivation (it was a mercantilist law) and political aims (it was an imperialist law)" (Bedjaoui, 1985). Hence, in the pre-colonial era, the comity of states bound by early international law was more or less a comity of European Christians, inspired by dreams of empire and conquest and emboldened by the prospects of looting the wealth of colonized peoples.[7] Accordingly, peoples and cultures outside the charmed circle of European Christianity were regarded as uncivilized savages in immediate need of civilization and enlightenment; even if the alleged civilizing mission inexorably led to series of genocide and dispossession of peoples.

The racist, imperialist, and religious basis of colonialism may easily be inferred from the fact that the weakest and the smallest European states or principalities were immune from colonization by members of the same Christian, European family. It is significant that in subsequent centuries when Adolf Hitler's Third Reich broke this taboo and sought to colonize and denigrate fellow Europeans, the colonial enterprise imploded. On the other hand, it is a remarkable contrast that while pre-Hitlerite European states were immune from colonialism by Europe, huge and sprawling empires, kingdoms, peoples and cultures in Africa, South America, *et cetera* were pillaged and colonized by Europe on the dubious doctrines of "discovery" and "*terra nullius*." Introduced in 1492, the European doctrine of discovery "opened the debate about European and Native American life ways and whether the original inhabitants of the Americas were a human or a subhuman species" (Nunes, 1995). Lamenting the colonial conquest of Africa, a newspaper noted in 1885 "the world has, perhaps, never witnessed a robbery on so large a scale" (Lagos Observer, 1885) and in the opinion of the newspaper, "this Christian business can only end, at no distant date, in the annihilation of the natives" (Mutua, 2000).

Colonialism was not simply a violent usurpation or overthrow of pre-existing political and economic structures of indigenous peoples across the world. Rather, the colonial enterprise was also an imposition of Eurocentric philosophies, values and world-view on indigenous and traditional peoples. In other words, beyond physical pillage and economic looting which the colonial project excelled at, it was a mechanism of cultural violence designed to remodel non-Western peoples and cultures in the image of Europe on the theory that indigenous peoples were sub-human. In the words of Makau Wa Mutua, "the 'Age of Empire' thus witnessed the forced assimilation of non-European peoples into international law, a regime of global governance that issued from European thought, history, culture, and experience" (Wa Mutua, 2000). Thus, while non-Western epistemologies, cultures, and value systems were dismissed as irrational, mystical, natural and undeveloped, Western norms of civilization, world-view, epistemology and culture were uniquely positioned as rational, empirical, and universal ideals and attainable by all regardless of differences in culture (Shiva, 1988).

Cultures or peoples that stood in the way of this monolithic conception were regarded as irritating relics of a primitive age. Echoing the dominant views of his day, famed British scientist, Robert Boyle, called for the annihilation of New England Indians for "their ridiculous notions about the workings of nature." In his view, the American natives were "a discouraging impediment to the empire of man over the inferior creatures of God" (Easlea, 1980). As Makau Wa Mutua notes, "within this logic, history is a linear, unidirectional progression with the 'superior' and 'scientific' Western civilization leading and paving the way for others to follow" (Wa Mutua, 2001). As non-Western epistemologies occupied the lower rung of the epistemological ladder designed by the colonial agenda, the path was laid towards the appropriation of indigenous peoples' knowledge. In this bizarre re-ordering of the world, Western forms of intellectual property protection became the only recognized and enforceable mechanisms for articulating and protecting intellectual property. International law proceeded as if there was no alternative framework for articulating and protecting intellectual property among the colonized peoples.

Although it is true that as early as the 16th century, Francisco de Vittoria had argued that the legal principles and concepts of indigenous peoples deserved respect, his views were in the minority.[8] Most of the colonizers of traditional and indigenous non-Caucasian societies and peoples held rather poor views of the natives whom they dismissed as "a barbarous race, possessing inferior rational capacities" (Margadant, 1980; Williamson 1992). In such intensely racist conception of peoples and knowledge frameworks, it was the opinion of some leading international law theoreticians of the day that "the Indians had an inalienable right to be slaves" (Margadent, 1980).

Perhaps a less abrasive but nonetheless insidious metamorphosis of the doctrine of racial inferiority of colonized peoples was the benevolent guardianship or "trusteeship doctrine" in international law that the colonial legal order imposed on the subjugated peoples and cultures. This attitude operated on the arrogant and bizarre notion that non-Europeans had the mental capacity of children and needed to be supervised and "assimilated" (Slater, 1994; Christina-Butera, 2001) into Western philosophy and culture. With particular reference to Africa, the Berlin Conference of 1885 at which Africa was carved up for the exploitation of Europe in the infamous "scramble for Africa" undertook to "watch over" Africans. Article VI of the Conference General Act proclaims the resolve of the parties "to bind themselves to watch over the preservation of the native tribes, and to care for the improvement of the conditions of their moral and material well being" (General Act of The Conference of Berlin, 1973). Considering the traditional dominance of Eurocentric attitudes and philosophies on international law, these policies and attitudes found expression in international law. Under the League of Nations, for example, the paternalistic, "superior race", took it upon itself to place indigenous peoples under its "guardianship." In fact, the last of such trust territories, New Guinea, gained independence in 1976.

Further, Article 12 of ILO Convention 107 of 1957 provides that "measures should be taken to facilitate the adaptation of workers belonging to the population concerned (traditional and indigenous peoples) to the concepts and methods of industrial relations in a modern society" (The Convention Concerning The Protection and Integration of Indigenous and other Tribal and Semi-Tribal Populations In Independent Countries, 1957). In effect, indigenous peoples were deemed morally, mentally inferior and primitive by the dominant legal and epistemological order. More importantly, in a patriarchal sense, they (peoples and cultures of the South) had become through an unilateral and self-imposed task of civilization and redemption, the "White Man's burden." Given the unclear difference, if any, between outright racial subjugation and a benevolent, patriarchal protection of these wards of imperial Europe, it is not surprising that until modern times, traditional knowledge frameworks and indigenous peoples' knowledge languished in the margins and merely served as objects of curiosity for anthropologists interested in the "exotic" and inquisitive of the "other."

In spite of the racist postulations of early European writers,[9] historiographers and apologists of the colonial enterprise, most scholars are now agreed that the history and scientific achievements of pre-colonial non-Western societies stretching across several millennia of profound and significant contributions to human civilization were undeserving of the centuries of marginalization and erasure (Davidson, 1967; Diop, 1987). From the awe-inspiring pyramids of Egypt to the epic lore of several West African communities (Okpewho, 1975), the contributions of pre-colonial non-Western communities have continued to amaze and inspire modern humanity.

With particular reference to plants and associated knowledge of the uses of plants, until recent times, the notion or philosophy that non-Western epistemologies were primitive, backward and unscientific afforded an attitudinal anchor, for the denigration and non-recognition of non-Western frameworks of knowledge. To worsen a miserable situation, a majority of the innovations and improvements in the farming fields and societies of traditional and indigenous societies are undertaken by women. Often, these contributions and innovations are ignored or ill appreciated in those cultural settings that ought to have championed the struggle for legitimating and recognizing those enormous contributions.[10] In effect, gender suppression within traditional and indigenous societies complicates our understanding of the denigration of traditional epistemology and innovative practices by the dominant Western narrative of science (Mgbeoji, 2005).

However, with reference to plants and associated traditional and indigenous knowledge, it is significant that virtually all crops of economic or social significance have their roots in the marginalized cultures and societies of the South. For example, the potato is indigenous to the Andes; maize is indigenous to Central America and rice originated from the Hindus Valley and West Africa. Indeed, of the 20 major food crops, none originated in North America or Australia and only two of the major food crops—rye and oats—originated in the Euro-Siberia area (Bosselman, 1995). Virtually all of the developed countries' foodstuffs originated in the tropical countries. Corn, rice, potatoes, sugar, citrus fruit, bananas, black peppers, nutmeg, pineapples, chocolate, coffee, vanilla, *et cetera*, all originated from the tropics. Jack Kloppenburg (1988) has captured this stark imbalance in plant distribution thus:

> [O]f crops of economic importance, only sunflowers, blueberries, cranberries, pecans, and the Jerusalem artichoke originated in what is now the United States and Canada. An all-American meal would be somewhat limited. Northern Europe's original genetic poverty is only slightly less striking: oats, rye, currants, and raspberries constitute the complement of major crops indigenous to that region. Australia has contributed nothing at all to the global ladder.

As various scholars have noted, the preponderance of plant genetic diversity amongst indigenous and traditional peoples is not solely a function of nature but an artificial result; a product of millennia of systematic breeding by traditional and indigenous peoples (Fowler, 1994). It is well known that cultural diversity correlates with genetic diversity. The greater the cultural diversity of local peoples, the more likely they are to breed plants for various cultural, economic, ecological and spiritual uses, thus multiplying the genetic diversity of plants (Isaac, 1970; Kameri-Mbote and Cullet, 1999). In dismissing the earlier notion that plant genetic diversity in the South is a freak of nature devoid of human mediation, Kloppenburg (1988) observes that:

> [I]n fact, the land races of the Third World, are most emphatically not simple products of nature. Traditional agriculturists have made very great advances in crop domesticity. Domesticated crops of species are frequently very different in form from their wild and weedy relations.

Lending additional weight to this fact, famous plant breeder, Norman Simmonds (1979) observed that "the total genetic change achieved by farmers over the millennia was far greater than that achieved by the last hundred or two years of more systematic science-based effort." In counselling against the hubris of the dominant Western scientific paradigm, Robert Leffel, the programme leader of the United States Department of Agriculture for Oil-Seed crops advised his fellow scientists that "in our modest moments, today's soybean breeders must admit that a more ancient society made the big accomplishment in soybean breeding and that we have merely fine-tuned the system to date" (Kloppenburg, 1988).

Notwithstanding these restatements that today reflect modern opinions on the intellectual worth of plant genetic improvement by local farmers and indigenous peoples, it is remarkable that for centuries, traditional and indigenous peoples' knowledge was denied legitimacy, scholarly recognition, and legal protection. The processes of de-legitimizing the intellectual inputs of local and traditional farmers and healers is a multi-pronged mechanism encompassing social repudiation of the intellectual merits of non-Western methods of inquiry, appropriation of indigenous peoples' knowledge, and the pervasive notion that indigenous peoples knowledge is unscientific. In addition, the sexism inherent in the masculinist origins of Western science has impacted negatively on the recognition of indigenous knowledge systems.

Western science has been promoted as a universal, absolute and de-cultured phenomenon. Conversely, non-Western epistemologies are exotic, culture-specific and ethnic manifestations lacking universal validity and relevance. This dichotomy is most apparent in the characterization of non-western epistemology as "ethnic" knowledge. In other words, Eurocentric conceptions of non-Western contributions to science have insistently sought to "ethnicize" and deny the empirical character and basis of non-Eurocentric narratives and scientific frameworks. The objective of this process, it would seem, has been to present non-Eurocentric paradigms of knowledge as "culture-specific" in contrast with Western science, which has always been presented, as "universal."

This powerful mis-characterization of non-Western epistemology constitutes the prevalent intellectual and philosophical context in which appropriation of indigenous peoples knowledge operates in modern times. Consequently, plants and associated knowledge of the uses thereof created or modified in traditional frameworks of knowledge have been systematically deemed incapable of protection as products of a credible systematized method of inquiry. Rather, they have permanently occupied the peripheral realms of "folk-knowledge," "ethno-botany," *et cetera*. For example, till date, it is virtually the tradition in modern discourse on traditional and indigenous peoples' knowledge to describe the study and knowledge of plants by scientists as "botany" while marginalizing non-Western knowledge and epistemology of plants as "ethno-botany."[11]

Non-Western empiricism has been largely misconstrued as an exercise in exoticism and a voyage to the worlds of charlatanism and quackery. In this sense, traditional knowledge is at best, a glorified form of "folk-lore" (Johnson, 1990) without empirical support while Western empiricism is unabashedly heralded as "scientific" and universal in character. In spite of this denigration of non-Western epistemologies, it is very striking that the products of traditional knowledge systems have defied the irrational badge of epistemological inferiority and ethnicity cast upon them by the dominant Western paradigm of knowledge. It is a well-known fact that the products of the so-called "ethnic" knowledge of traditional peoples and cultures, which ought to have been "culture-specific", command global validity irrespective of where such products are applied or used. The examples are legion but a few will suffice. For instance, the serpent tree which has been in use in India for thousands of years as an herbal remedy for some types of mental illnesses is used today to make drugs such as *reserpine* for epilepsy and high blood pressure.

Hallucinogenic and pain relieving plants such as *papavar somniferum* and *erythroxylon coca*[12] which were used as such by Indians are still used in modern days to produce morphine and cocaine (Sinha, 1996). Tobacco which was first used by the Arawak Indians of North America is today a globally known and distributed drug. Of course, given the widespread use of these plants today, especially in Europe and the United States, it would be absurd to suggest that those "ethnic" plants and their products such as morphine, cocaine or cigarettes express their efficacy as drugs only on people living in traditional "ethnic" cultural frameworks.

As already mentioned, quinine, which is used all over the world for the treatment of malaria is also an "ethnic" product of the traditional knowledge of the uses of the tree- *cinchona officinalis*. It has also been proven that the "ethnic" plant *withania somnifera* has an anti-tumour property; *gymnema sylvestre* has curative properties against diabetes and *centella asiatica* is anti-leprotic. In addition, *embelia ribes* is good as antihelminthic and *messua ferrae* is good for respiratory disorders. *Carum coticum* lowers blood pressure. The list is virtually endless and all these herbs have pharmacological properties consistent with the original insights developed by the "primitive" peoples of the "ethnic" cultures that developed them (Sinha, 1996).[13]

Throughout colonial history and even in modern times, the stupendous knowledge of plants possessed by the so-called primitive and tribal peoples of the forests and jungles of the Third World have sustained many industries and civilizations. The World Resources Institute estimates that "Indians dwelling in the Amazon Basin make use of some 1,300 medicinal plants, including antibiotics, narcotics, abortifacients, contraceptives, anti-diarrheal agents, fungicides, muscle relaxants and many others" (Miller et al., 1991). Similarly, recent studies show that the Epugao of Luzon in the Philippines can identify 200 varieties of sweet potatoes and the Andean farmers cultivate thousands of clones of potatoes, more than 1000 of which have names.

The economic significance of indigenous and traditional knowledge of plants can hardly be overstated. Over a quarter of modern drugs prescribed all over the world are directly derived from plants and most of them are products of the much denigrated "ethnic" and "primitive" traditional knowledge of the uses of plants. Over 80% of peoples in the South rely on plants for their medicinal supplies, thus, justifying the efforts of the World Health Organization (WHO) to incorporate traditional healers in the dispensation of health care to billions of people all over the world (Meyer et al., 1981).

The absurdity in the characterization of plants developed and or improved by traditional and indigenous cultures operating within their own epistemological frameworks and thus ostensibly lacking in universal empirical validity is perhaps best demonstrated in the global use of food crops hitherto regarded as ethnic crops. Today, virtually every urban neighbourhood in the world has a coffee shop. Yet, coffee is an "ethnic" crop and beverage which originated from the tribal mountains of Ethiopia and Kenya in Africa. Columbus brought maize into Spain in 1496 and today it is a major component of the diets of several peoples of the world, especially, in East Africa. Similarly, the popular drink Coca-Cola which is probably the most universally recognizable "face" of global commerce is partly derived from two "ethnic" crops: the West African Kola nut and Gum Arabic from Sudan. Popular food crops and drinks aside, before the settling of Europeans in North America, native Shoshone Indian women of Nevada chewed stone seed for birth control purposes.[14] Western scientists initially scoffed at the idea of birth control through stone seed. Later, Western science confirmed that stone seed contains estrogen, which regulates ovulation. Today, virtually all birth control pills contain estrogen. Indeed, the revolutionary implications on human rights and the empowerment of women wrought by this "culture-specific", "folk-lore" is so well known.

INTERNATIONAL LAW AND THE RE-EMERGENCE OF INDIGENOUS KNOWLEDGE SYSTEMS

Over the past couple of decades, the indigenous question has traversed the international law landscape, moving from the larger question of political self-determination for indigenous peoples across the world to specific issues of cultural, epistemic and spiritual identity (Wiessner and Battiste, 2000). With this shift and metamorphoses in movement, it is increasingly recognized in many quarters that non-Western epistemologies (Ezeabasili, 1977) are indeed verifiable forms of knowledge that should be protected with new, hybrid, or indigenous autochthonous forms of intellectual property regimes (Coombe, 2001). This change in juridical position may not be unconnected with the end of formal colonialism and the unrelenting agitation by indigenous and traditional peoples (with the support of activists and non-governmental organizations and sympathizers in the North). Another possible factor is the increasing recognition of the holistic nature of indigenous knowledge. It is perhaps in recognition of the narrowness of Western empiricism and its troubling neglect or denial of the multiple dimensions of knowledge that Western epistemology is increasingly becoming multi-disciplinary in its approach to the articulation, acquisition, and deployment of knowledge.

The emergence of indigenous knowledge as a protectable and credible method of inquiry[15] probably started after the end of colonialism and the attendant re-evaluation of colonial biases against indigenous peoples' and their knowledge systems. Given the inter-penetration of knowledge frameworks (Edsman, 1962; Lommel, 1967; Loudon, 1976) and the complementarity of such knowledge frameworks,[16] an antagonistic state of affairs seems unhelpful. Drawing support from the emerging normative shift in world relations, anthropologists and bioprospectors gradually dismantled earlier erroneous notions that traditional and indigenous knowledge frameworks lacked empiricism. As one of such researchers concluded:

> [T]he impressive knowledge of the Native American peoples about a wide variety of natural phenomena is not however accidental, nor has its acquisition been haphazard. It is based on generations of systematic inquiry. It is the accumulation of and transmittal of repeated observations, experiments and conclusions. Some of the elements of the scientific method were inherent in their processes (Landy, 1977).

As Michael Balick further argues a "deeper reluctance to explore indigenous knowledge systems may be attributed to cultural prejudice dating to the years when the Western powers reigned over colonies. If cultural biases are eschewed indigenous traditions and science are epistemologically closer to each other than Westerners might assume."[17]

Quoting F.S.C. Nortrop, Peter Morley also affirms:

> [O]ne must seriously ask oneself whether superstition and myth, in the derogatory or non-scientific connotations of these words, are not due to our judging a given people from our conceptual standpoint, rather than theirs...when the trouble was taken to find their concepts, then it became evident that everything made sense and that their behaviour and cultural norms followed as naturally and consistently from their particular categories of natural experience as ours do from our own. I believe it is just as much an error to suppose that there were no people anywhere who insisted on empirically, and hence scientifically, verified basic concepts before Galileo. Prevalent as the latter is, it is nonetheless nonsense.[18]

Notwithstanding the change in the dominant scientific paradigm, changes in the juridical character of traditional peoples' knowledge systems at the international level have had to await the demise and dismantling of formal and substantive colonialism.¹⁹ Notwithstanding Prime Minister Churchill's bifurcated comprehension of the normative scope of the Atlantic Charter, colonial subjects were inspired by the Charter and it was becoming clear that formal colonialism was at its eventide. A wind of change was sweeping away the age of colonialism and with, the implied, if not explicit subjugation of indigenous peoples' knowledge. This normative tide, particularly in respect of indigenous peoples and their cultures soon swelled the juridical banks as expressed in the Universal Declaration of Human Rights adopted and proclaimed at the General Assembly of the United Nations on 10 December 1948.

Article 27 thereof captures the essence of this normative revolution. It provides that "every one has the right to freely participate in the cultural life of the community, to enjoy the arts and to share in scientific advancement and its benefits" (Universal Declaration of Human Rights, 1948). In addition, Article 27(2) provides that "everyone has the right to the protection of the moral and material interests resulting from any scientific, literary, artistic production of which he is the author." Similarly, article 27 of the International Covenant on Civil and Political Rights provides that:

> [I]n those states in which ethnic, religious or linguistic minorities exist, persons belonging to such minorities shall not be denied the right, in community with the other members of their group to enjoy their own culture, to profess and practice their own religion, or to use their own language (U.N.T.S., 1966).

It is arguable that this normative shift, a post-colonial development, gained momentum with the comments of Halfdan Mahler, the Director-General of the World Health Organization (WHO) in 1977. As he pointed out, "let us not be in doubt: modern medicine has a great deal still to learn from the collector of herbs" (Ackerknecht, 1991). Mahler's summons granted official *imprimatur* to non-Western traditional knowledge of the medicinal uses of herbs, a system that has served billions of people for millennia.

These attitudinal, perhaps, normative shifts in the conception of indigenous and traditional peoples laid the foundation for the jettisoning of previous provisions and attitudes in international law and institutions which had construed or perceived traditional or indigenous peoples' knowledge systems as primitive and superstitious (McClellan, 2001). Hence, the doctrine of assimilation of non-Western cultures into the dominant Euro-centric paradigm came under scholarly and cultural attacks from both the liberal wings of the academe, indigenous peoples' forums, and their army of sympathizers in the North.

Starting in the 70s, many indigenous peoples groups and activist scholars realized that the creation of International Agricultural Research Centres (IARCs) as the controlling agencies for international gene banks tended to serve the interests on industrialized states. Indeed, the concept of IARCs was conceived in a meeting in 1941 between Vice-President Wallace of the United States and the President of the

Rockefeller Foundation, Raymond Fosdick. Vice-President Wallace (he was to be highly involved with Pioneer Hi-Bred Inc.) believed that a program of agricultural development aimed at Latin America in general and Mexico in particular would have both political and economic benefits for the United States (Kloppenburg, 1988). By 1943, the Rockefeller Foundation started its Mexican Agricultural Program by which the collection of indigenous germ plasm was an important component of the program. The IARC project was timed to mitigate the losses which North American and European states would suffer in terms of loss of access to Third World genetic resources. As William Lesser (1997) has shrewdly observed, with the "decline of the empire system...governments lacked the military presence and legal authority to compel sovereign nations to yield valuable germ plasm". Thus, the IARC projects probably marked the beginnings of a post-colonial institutionalization of the global network for the appropriation of the South's germ plasm on the ostensible grounds of fostering research in agricultural crops.

The location of the IARCs in the various countries of the South was not a product of fortuitous occurrences or a series of coincidences; rather, the IARCs were established in regions with known phenomenal stock of indigenous germ plasm, that is, the so-called Vavilov centres. For example, the International Rice Research Institute (IRRI) is located in the Philippines; the International Center for Agricultural Research in Dry Areas (ICARDA) is located in Syria; the West African Rice Development Association (WARDA) is located in Liberia; the International Potato Center in Peru. Other IARCs include the International Institute of Tropical Agriculture (IITA) in Nigeria, the International Crops Research Institute for the Semi-Arid Tropics (ICRISAT) in India, and the International Maize and Wheat Improvement Center (CIMMYT) in Mexico (Kloppenburg, 1988). Thus, for each crop of global significance, such as corn, wheat, rice, and other cereals, the IARCs were established at the respective centre of diversity and origin and then the plant germ plasm was funneled to the North. In little time, an enormous quantity, quality, and diversity of plant germ plasm had been collected

Through the IARC projects, plant germ plasm collected in Third World countries were released into the farm fields of the United States and Europe without any economic benefits for the South or recognition of the intellectual property rights of the local farmers, particularly women, who had spent millennia (and still continue to do so) in improving the plant germ plasm in question. Interestingly, while the genetic resources were collected from Third World states, storage of the germpasm took place in industrialized states of the North, especially, the United States. By 1956, the US constructed the massive and sophisticated National Seed Storage Laboratory (NSSL) at Fort Collins, Colorado, United States. It should be noted that the plant germ plasm could easily have been stored in their countries of origin but this never happened. Over 80 percent of all economically useful plant germ plasm (accession) and varieties indigenous to the South collected by the IARCs are stored in gene banks located in the industrialized states of North America and Europe. The Food and Agriculture Organization (FAO) has identified over 1200 plant genetic resource collections worldwide, held in more than 160 countries and territories

(UNEP, 1994). Overall, governments hold 83% of the accessions, the International Agricultural Research Centres (IARCs) 11%, and the private sector 1.27%. The IARC collections contain about 35% of the unique samples, making them probably "the world's most significant collection" (Lesser, 1997) of Southern plant germ plasm. As Naomi Roht-Ariazza notes, "most genetic materials collected in Southern countries-68% of all crop seed, 85% of all livestock, and 86% of microbial culture collections are held at the IARCs or in Northern countries" (Roht-Arriaza, 1996) Consequently,

> [T]he advanced capitalist nations, though poor in naturally occurring plant genetic diversity, are as rich in "banked" germ plasm as the developing nations of the Third World. Indeed in a number of crops (wheat, barley, food legumes, potato) the advanced capitalist nations possess more stored germ plasm accessions than do those nations that are the regions of natural diversity for the crop (Kloppenburg, 1988).

Without question, the activities of the IARCs constitute the most far-reaching institutional appropriation of traditional and indigenous owned plant germ plasm in recent times. The IARCs constituted an efficient *vehicle for the efficient extraction of plant genetic resources from the Third World and their transfer to the gene banks of Europe, North America, and Japan* (Kloppenburg, 1988). It is equally noteworthy that storage of the germ plasm *ex situ* in the North sometimes rendered such germ plasm inaccessible to the original providers from the South (Wilkes, 1977). The IARCs were placed under the control of the CGIAR. The dangers of excessive concentration of power in the hands of the gene banks located in a few powerful industrialized states led an FAO-organized conference to conclude that a coordinated global program of collection and conservation was necessary to ensure that the essential raw materials of plant improvement were not lost. In 1972 the UN Conference on the Human Environment in Stockholm issued a resolution calling for an international program to preserve the germ plasm of tropical crops. In consequence, the Beltsville Conference in Maryland in 1972 recommended the establishment of the International Board for Plant Genetic Resources (IBPGR). The significant fact here is that instead of locating such an international programme under the FAO which is an agency of the United Nations, the Northern-controlled CGIAR argued that it [the CGIAR] should be designated the research arm of world agricultural development.

The anomalous compromise in this situation was the creation by the CGIAR in 1974 of the International Board for Plant Genetic Resources, (IBPGR). The IBPGR was under the political and financial control of the industrialized states of the North. The loyalty of the IARCs to the North is also reflected in the ideological and political stance of the IARCs in the non-release of plant germ plasm to states perceived by it to be ideological enemies, especially during the Cold War era. More worrisome, some powerful industrialized states began to make legal claims to ownership of the accessions held by the gene banks. In some cases, some of the gene banks, for political or economic reasons, sometimes refused to release or grant free access to the stored germ plasm to other states even when those states whose requests were rejected were the original donors of the plant accession in question. For example, the gene banks located in the United States have been known several times to refuse free access to requests from original donors such as Afghanistan, Albania, Cuba, Libya, Nicaragua, and the defunct Soviet Union (Kloppenburg, 1988).

As ownership and intellectual property rights began to be asserted over genetic resources held in trust by the gene banks, the global South began to murmur their dissent. The North dismissed the arguments of the South and observed that there were no legal barriers stopping the South from "improving" the plant germ plasm itself and re-selling it in the global market as it wished. The discontent of the South was soon to find expression in the framework of the FAO and ultimately led to an aggressive re-assertion of national and state sovereignty over plant genetic resources. Ultimately, the juridical solution was the emergence of an international legal order that recognizes the peculiar character of plant genetic resources as both a national and global asset.

More importantly, the emerging debate on what constituted "improvement" of plant genetic resources revealed a huge epistemic and cultural divide between the South and the North. While the North had no difficulty in construing its varieties as scientific products deserving of the greatest respect, there was pervasive notion that the accessions held in trust were mere raw materials. The dichotomy sought to nullify millennia of efforts by local farmers, breeders and healers across the world in improving and conserving biodiversity (Tillford, 1998). Various undertakings negotiated by the FAO in the early 1980s brought the issue of non-recognition of the intellectual inputs of the Third World to biodiversity improvement to a head (Vogel, 1994).

The differences between the North and the South also focused on the legal status of the gene banks and control of the IARCs. Given that these gene-banks hold at least one-third of the unduplicated samples of the world's plant germ plasm, it is not surprising that an acrimonious legal debate ensued between the North and the South as to the exact legal status of the IARCs and the plant accessions held in those gene banks. However, the CIMMYT—International Maize and Wheat Improvement Centre, and IRRI-International Rice Research Institute—seemed to have clearer policies on ownership of germ plasm in storage. With respect to the other IARCs, the position was far from clear. A nebulous legal opinion by the FAO in 1986, equivocated that the CGIAR gene banks:

> [E]xisted in a unique world between national and international law. They are not created by a formal treaty concluded among States or other international legal persons, and their activities are not directed by States or such other international legal persons. The gene banks maintained by the IARCS are neither under the control of any given State or national authority, nor in the private sector. Their status is, in fact, *sui generic* (Tillford, 1998).

Another legal opinion of the FAO in 1987 hardly added clarity to the question. It intoned that:

> [O]wnership of genetic material held in government gene banks or those of public institutions was in most cases, for practical purposes considered to be vested in the State in which these gene banks are located. However, for material held in the International Agricultural Research Centres, the legal position was unclear (UNEP, 1994).

This obfuscation fuelled the debate for a clarification of the legal status of genetic material held by the IARCs. More importantly, the debate underscored the need to recognize and protect indigenous peoples knowledge. Part of the suggestions made to resolve the impasse was the creation or recognition of the Farmers' Right in various FAO Agreements and Undertakings. At a review of the FAO International Undertaking of 1983 in Leipzig, Germany, a global plan of action was adopted. The Global Plan of Action (although non-binding) is perhaps one of the earliest international documents recognizing the intellectual input of local farmers in the improvement and conservation of plant germ plasm. However, its platitudinous and patronizing praises offered little comfort to indigenous/local farmers.

The normative changes adumbrated by the FAO Undertakings on the recognition of indigenous peoples' intellectual contributions to biodiversity were subsequently inscribed in international legal instruments such as the ILO Convention 169 of 1989. The preamble to the Convention provides the key elements of this normative change. It notes, *inter alia*, that:

> [C]onsidering that the developments which have taken place in international law since 1957, as well as developments in the situation of indigenous and tribal peoples in all regions of the world, have made it *appropriate to adopt new international standards on the subject... with a view to removing the assimilationist orientation of the earlier standards...* and recognizing the aspirations of these peoples to exercise control over their own institutions, ways of life, and economic development and to maintain and develop their identities, languages, and religions, within the frameworks of States in which they live, and...
>
> Noting that in many parts of the world these peoples are unable to enjoy their fundamental human rights to the same degree as the rest of the population of the States in which they live, and that their laws, values, customs and perspectives have often been eroded, and...
>
> calling attention to the distinctive contributions of indigenous and tribal peoples to the cultural diversity and social and ecological harmony of humankind and to international co-operation and understanding... (International Labour Organization, 1989)

These preambular contributions should not be discounted because introductory clauses in treaties are legitimate sources of international law (Hall, 1977). Moreover, the Vienna Convention on The Law of Treaties, which is a codification of the customary principles of general international law, obliges states to take preambular provisions into account in the context of construing the substantive provisions of treaties. Beyond the preamble of the ILO Convention 169 of 1989, articles 4 and 5 of the Convention oblige states to take special measures to protect the integrity of the labour, cultures and values of indigenous and traditional peoples. In general, ILO Convention No. 169 offers a legal basis for the protection of nearly all aspects of indigenous peoples' knowledge.

Without a doubt, the most significant and celebrated international law instrument on the recognition and protection of indigenous and traditional knowledge is the 1993 Convention on Biological Diversity (CBD). Rather than leave indigenous peoples' knowledge in the doldrums or peripheries of neglect by states, Article 8(j) of the CBD requires state-parties to:

> [R]espect, preserve and maintain knowledge, innovations and practices of indigenous and local communities embodying traditional lifestyles relevant for the conservation and sustainable use of biological diversity and promote their wider application with the approval and involvement of the holders of such knowledge, innovations and practices and encourage the equitable sharing of the benefits arising from the utilization of such knowledge, innovations and practices.[20]

This provision does not only recognize that indigenous and traditional societies have extensive knowledge of their surroundings but also accepts that such communities has and will continue to give critical clues to other scientists working in various areas such as agriculture, medicine, and industry. Implicit in the recognition of the desirability of sharing benefits with these communities is the juridical validation of the scientific nature of the knowledge systems and techniques of traditional and indigenous peoples. Article 18(4) further requires state-parties to "encourage and develop methods of cooperation for the development and use of technologies, including indigenous and traditional technologies, in pursuance of the objectives of" the CBD.

It would be incorrect to proceed as if the CBD provision on the need for state recognition and protection of indigenous and traditional peoples' knowledge systems in unprecedented in international law. Indeed, the provisions of the Food and Agriculture Organization (FAO) Undertaking of 1989 had adumbrated Article 8(j) of the CBD. Resolution 5/89 of the FAO which defines Farmers' Rights, articulates the nature of indigenous and traditional peoples rights over plant genetic resources as "rights arising from the past, present and future contributions of farmers in conserving, improving, and making available plant genetic resources, particularly, those in centres of diversity/origin."[21]

The emergence in 2001 of the FAO Treaty on Plant Genetic Resources has not silenced debates on the aforementioned issues. The centerpiece of the treaty is its multilateral system for access and benefit sharing. It is generally acknowledged that access to the Third World's wide genetic base of plants will allow the further development of improved varieties. The sticking point, however, has remained the lack of compensation to and recognition of the intellectual contributions of the Third World peoples and cultures towards the improvement and conservation of plant varieties. The FAO treaty is in many respects a comprehensive instrument and thus its Article 9 would seem to remedy the extant shortfall in international law pertaining to the denial of indigenous contributions to plant development. Article 9 of the treaty states that parties are to:

> [R]ecognize the enormous contribution that the local and indigenous communities and farmers of all regions of the world, particularly those in the centres of origin and crop diversity, have made and will continue to make for the conservation and development of plant genetic resources which constitute the basis of food and agricultural production throughout the world.

In the context of the treaty, recognition of indigenous intellectual contribution to plant conservation and development is by way of an improved "farmer's rights." It should be noted that "farmers' rights" was a concept originally introduced into international discourse in 1988, when the Colorado-based Keystone Center mediated in the dispute between several states and the center under the aegis of the Keystone International Dialogue on Plant Genetic Resources. The Keystone mediation accepted the concept of farmers' rights proposed by Mexico in exchange for a regime of free access to Third World germplasm under the defunct 1989 FAO Undertaking As articulated in the FAO Undertaking of 1989 as per Resolution 5/89, Farmers' rights are defined as:

> [R]ights arising from the *past, present and future contributions of farmers in conserving, improving and making available plant genetic resources, particularly those in centres of origin/diversity. These rights are vested in the international community as trustee* for present and future generations of farmers for the purpose of ensuring full benefits to farmers and supporting the continuation of their contributions, as well as of the overall purposes of the International Undertaking (FAO Treaty).

It is instructive that under the concept of Farmers' Rights as articulated under the 1989 Undertaking, "future" intellectual exertions and input of farmers from the South, in improving plant germ plasm are vested in the "international community." In sharp contrast, intellectual exertions by "scientists" in the laboratories of the multinational seed companies of the North are construed as private property and secured with patents and plant breeders' rights. While these international instruments are non-binding agreements are now dated and obsolete, they provide a useful sub-text towards a nuanced understanding of the politics and economics of the struggle for control of plants through the mechanism of patents, legal concepts, and international institutions.

But do these provisions create a solid and enforceable regime for the protection of indigenous and traditional peoples' knowledge (Santiago et al., 2004)? It is very arguable that the series of non-binding resolutions adopted by the FAO in respect of indigenous and traditional peoples' contributions to plant genetic improvement paid lip-service to the need for juridical respect for and protection of indigenous peoples' knowledge. It must be noted, however, that the recently concluded FAO *International Treaty on Plant Genetic Resources for Food and Agriculture* is a bold departure from the FAO resolutions. It is an innovative juridical provision on the need for the protection and promotion of indigenous peoples knowledge.

The FAO (2001) treaty which has already entered into force, obliges member-states and parties to recognize and protect the

> [E]normous contribution that the local and indigenous communities and farmers of all regions of the world, particularly, those in the centres of origin and diversity (the South) have made and will continue to make for the conservation and development of plant genetic resources which constitute the basis of food and agriculture production throughout the world.

In addition, Article 5 of the FAO treaty specifically requests state parties to take appropriate measures toward the protection of indigenous and traditional peoples knowledge. The FAO treaty makes three substantive contributions to the concept of farmers' rights. These are (1) the protection of relevant traditional knowledge (echoing Article 8(j) of the CBD); (2) recognition of the right of farmer's to participate equitably in sharing benefits arising from the utilization of plant genetic resources for food and agriculture; and (3) recognition of the right of farmers to participate in making decisions at national levels. These are significant improvements on and contributions to the jurisprudence on protection of indigenous peoples knowledge. In fact, the FAO treaty, in this respect goes further than Article 8(j) of the CBD. However, Article 9(3) of the treaty specifies that:

> [N]othing in this Article shall be interpreted to limit any rights that farmers have to save, use, exchange and sell farm-saved seed/propagating material, subject to national law and as appropriate.

While this provision is seemingly neutral, given the limitation of responsibility for implementation to national governments, it is arguable that this is a loss for farmer's, especially, indigenous and traditional farmers. Another important aspect of the FAO Treaty on Plants is the creation of a limited commons in certain food-related plant genetic resources. Thus, despite its multilateral system, the scope of the treaty is limited to a list of crops. The reason for this is that certain countries generally rich in biodiversity-even if they are not equally in plant genetic resources for food and agriculture-wanted to limit the application of the multilateral system, thereby leaving some room or potential for future bilateral arrangements. Similarly, it was agreed by virtue of Article 12(3)(a) that material made available through the FAO multilateral system should be "provided solely for the purpose of utilization and conservation for research, breeding, and training related to food and agriculture." In effect, chemical, pharmaceutical and/or other industrial uses beyond food and animal feed are excluded.

Clearly, while these changes in international law constitute a departure from the era when traditional knowledge systems were regarded or construed as "folk-knowledge" (Hardy, 1994; Adair, 1997) or barely disguised charlatanism, they do not constitute a solid basis for the protection of indigenous and traditional peoples knowledge. It is remarkable that while Eurocentric forms of intellectual property such as patents, computer software, and copyrights enjoy specific, rigourous and robust legal protection, scholars of indigenous peoples knowledge have to scour and scrounge an array of non-binding and ambiguous instruments to conjure schemes for the legal protection of indigenous peoples knowledge in international law.

In recent times, there has been a deluge of hortatory ethical declarations by several institutions, research centres, and organizations seeking to regulate access to, acquisition and use of indigenous peoples knowledge. For example, in 1988, the International Society of Ethnobiology adopted the Declaration of Belem. The Declaration marked the first time an international scientific organization recognized a basic obligation that "procedures developed to compensate native peoples for the utilization of their knowledge and their biological resources." Similarly, following on the Declaration of Belem (1988), the International Society of Ethnobiology adopted a code of ethics at its annual general meeting held at Whakatane, Aotearoa/New Zealand in 1998. Not to be left out, Shaman Pharmaceuticals, the American National Institutes of Health and several other institutions have adopted various ethical guidelines on access to and use of indigenous peoples knowledge. But do these ethical guidelines by well meaning institutions rise to the level of enforceable legal obligations in international law? The answer is, of course, a resounding no.

The difference in the way indigenous peoples' knowledge has been treated has raised significant questions about the status of indigenous peoples' knowledge in modern international law (Drahos, 1997). While most of the leading scholars on the subject posit that indigenous peoples knowledge has been recognized in international law (Coombe, 2001), it must be borne in mind that mere recognition is not coterminous or coeval with protection. A thing may be recognized and yet lack legal protection. Nowhere in the 1994 WTO Agreement on Trade-related Aspects of Intellectual Property (TRIPs Agreement) is any specific provision made for the protection of indigenous peoples knowledge. Yet, the TRIPs Agreement makes ample and robust provisions for the protection of Western denominated forms of intellectual property. If the recognition of indigenous peoples knowledge in international law must move from tokenistic and platitudinous verbiage, international law must come to terms with the fact that recognition of indigenous peoples knowledge without protective enforcement is ineffectual (Coombe, 1998).

But what have international legal frameworks done to protect indigenous peoples knowledge? In the first place, it has to be noted that "in the last decade or so, it has been acknowledged that knowledge protection mechanisms exist in every culture, and that such mechanisms need not necessarily be in the mould of Western intellectual property rights" (Oguamanam, 2004). The implication of this fact is that there must be a cross-cultural dialogue that would produce understandings of how to articulate and give legal force to the protection of existing indigenous knowledge protection protocols.

The Convention on Biological Diversity through its Inter-Sessional Working Group on Article 8(j) of the CBD has done some work on this cross-cultural dialogue. But like most issues affecting indigenous peoples, more talks; meetings and toothless declarations would take the place of effective and enforceable legal mechanisms for the protection of indigenous peoples knowledge. Similarly, the Global Intellectual Property Issues Division (GIPI) of the World Intellectual Property Organization (WIPO) has been set up to look into, *inter alia*, the protection of traditional knowledge, innovations, and creativity. Professor Rosemary Coombe (1998) has argued that "WIPO and CBD have recognized that indigenous customary law has to be respected when considering the use of traditional knowledge and that indigenous customary law principles provide legitimate juridical resources for a consideration of alternative forms and norms of property." It is within this context that WIPO has conducted global fact-finding missions that explored the intellectual property needs of indigenous peoples across the globe.

The implicit lesson thus is that international law may well recognize indigenous peoples knowledge but the protection of such knowledge would have to be inaugurated and primarily instituted in the domestic legal order. As the details of the emerging regime are being worked out at various domestic levels, particularly, in respect of dominant intellectual property regimes that have facilitated the appropriation of indigenous knowledge (Mgbeoji, 2001b), it must be remembered that centuries of neglect and exploitation cannot be remedied overnight, particularly, at the domestic levels.

Ultimately, it is at the domestic levels that the normative shifts occurring at the international levels would have to be worked out. Although international law recognizes two mechanisms for the domestic enforcement of international obligations, it seems that in the final analysis, the self-interests and value systems of state-parties to these new international treaties would play pivotal roles in local attempts at protecting indigenous peoples' knowledge. Given the multiplicity of autochthonous forms and norms of intellectual property among indigenous peoples, it would be unrealistic to expect a monolithic intellectual property regime that satisfies the plurality of indigenous peoples. Even though Article 26 of the Vienna Convention on the Law of Treaties provides that every treaty is binding upon the parties to it and must be performed in good faith, it is probable that domestic politics, capabilities and priorities would significantly affect how the promises of the new international regime on indigenous peoples' knowledge systems are fulfilled; (Seidl-Hohenveldern, 1963) more importantly, where such treaty obligations are notoriously couched in vague and hortatory language.

CONCLUSION

In the preceding pages, this chapter has argued that appropriation of indigenous peoples' knowledge is rooted in the colonial assault on indigenous and traditional knowledge systems. In effect, the phenomenon of appropriation and denigration of indigenous knowledge systems cannot be divorced from the cultural and economic underpinnings of the colonialist agenda. Accordingly, modern scholarship on the issue of devising viable regimes for the protection and promotion of indigenous peoples' knowledge of the uses of plants would have to re-examine the historical causes of modern biopiracy.

Further, the chapter has attempted to show that the mis-characterization of indigenous and traditional peoples knowledge as superstition and quackery is an unfair generalized castigation of alternate perspectives on the windows of the world. Furthermore, it has been demonstrated that the emergence of indigenous knowledge systems as legitimate and protectable systems of knowledge may be linked to the decolonisation process of the last century. Although modern international law recognizes indigenous and traditional peoples' knowledge, in the final analysis, however, it would take more than international law provisions to dismantle centuries of denial and erasure of the profound contributions of non-Western epistemologies.[22] A lot of work must de done at the various domestic levels. If indigenous and traditional peoples knowledge must be rescued from its blighted existence in the world of the dead and the nearly dead, states must in good faith discharge the enormous moral and legal debt they owe to indigenous peoples. The first reasonable steps in this regard would be to explore the juridical resources already recognized by indigenous peoples in their daily production, use, sharing, and propagation, of knowledge.

NOTES

[1] There are textual differences in the definition of indigenous peoples in international law (see ILO Convention (1989), Indigenous and Tribal Peoples, Martinez (1986). Interestingly, most of the definitions tend to have more relevance to the colonial histories and exploitation of the native populations of the Americas and Oceania.

[2] In this chapter, the term "South" refers to the gene-rich, ecologically diverse states of Africa, South America, Asia (excluding Japan) and Oceania. These states are also referred to or described as "less developed", "developing", or the "Third World" countries of the world. Considering the similar experiences of indigenous cultures and communities of North America, it makes sense to describe these minority and marginalized peoples of North America (Canada, the United States and Mexico) as part of the South (see Adams, 1993). However, it must be noted that the concept of "South" does not create a homogenous or monolithic structure or culture of peoples. This concept is only used for the ease and convenience of analysis (see Langley, 1981).

[3] On the other hand, the term "North" as deployed in this chapter refers to the states of Europe, North America, and New Zealand, Australia, and Japan. They are also variously described as the "rich", "industrialized", "developed" or "advantaged" states of the world. Generally speaking, these countries tend to share a similarity of relative poverty of naturally occurring or indigenous biological diversity.

[4] For an attempt to differentiate Western empiricism and indigenous peoples' knowledge see Martha Johnson (1992).

[5] In his material dialectics argument, Marx acutely observed that:

> [T]he discovery of gold and silver in America, the extirpation, enslavement and entombment in mines of the indigenous population of that continent, the beginnings of the conquest and plunder of India, and the conversion of Africa into a preserve for the commercial hunting of black skins, are all things which characterize the dawn of the era of capitalist production. These idyllic proceedings are the chief moments of primitive accumulation ...These treasures captured outside Europe by undisguised looting, enslavement, and murder flowed back to the mother-country and were turned into capital there (see Marx, 1977).

[6] Quinine, is the well known anti-malarial which comes from the bark of the Peruvian *cinchona* tree. For a detailed account of the development and political economy of quinine in the colonial age see Lama (2000).

[7] For an examination of the ethnic character of international law, see Jessup (1973), Coombe (1995).

[8] De Vittoria was supported by others like Bartolome las Casas who opined that the natives were creatures of God and endowed with the same rational capacities as the invading Europeans. Similarly, Pope Paul III in the Papal Bull of 1537 clearly noted that the Indian were human beings with the same rational abilities as the Europeans (see Cohen, 1942).

[9] It is a well-known fact that some European writers like Immanuel Kant, David Hume, *et cetera* had argued that Africans lacked a history and civilization.

[10] It is equally true that even in the dominant Western scientific paradigm the scientific and technological achievements of women are often uncelebrated. For a compendium and analysis of ground-breaking inventions by women and which have not been duly recognized in the popular narratives (see Stanley, 1993).

[11] The term "ethno-botany" was coined by the American botanist, John W. Harshberger in 1895 to describe studies of plants used by "primitive" and aboriginal peoples.

[12] According to eminent anthropologist, Prof. Vogel, until 1884 when Kohler distilled cocaine from coca leaves, accounts by native Americans of the immense powers of the coca leaves were dismissed. Physicians using coca to relieve pain of patients were made subjects of ridicule, as being incapable of judging a remedy's qualities. Pharmacists making preparations of the drug were looked upon askance; as being concerned in a fraud, while the natives who employed it in their daily life, as well as the travelers who were impressed by what they had observed of its effects, were regarded as involved in ignorance, or imbued with superstitious imaginings (see Vogel, 1992).

[13] For example, years before James Linds' "experiment" on scurvy, the American Indians had known the cause and cure of scurvy.

[14] Reflecting the general Euro-centric off-hand dismissal of traditional medicinal knowledge in those times, the scientist Norman Hines as late as 1936 dismissed the claims of the Shoshone women as useless. Without deigning to verify their claims, he maintained that "no drug has yet been discovered which, when taken by mouth, will induce temporary sterility" (see Johnson, 1990). However, other doctors were not so blinded by prejudice. Dr. Benjamin Barton of the medical faculty of the University of Pennsylvania declared in 1798 that the Indians had knowledge of some of the most inestimable medicines. He wrote a treatise on sixty indigenous plant remedies and fourteen of his students wrote dissertations on native remedies and practice. It is equally interesting to note that the discoverer of insulin, Dr. Frederick Banting was tremendously inspired by the Indians of British Columbia who used extracts from the devil's club-*fatsis horrida* for the treatment of diabetes (Johnson, 1990).

[15] William Lesser has also reported that recent studies clearly show that indigenous farmers and breeders were clearly aware of the formalized findings of Mendel on genetic traits and breeding. According to the Crucible Group, traditional "farmer's fields and forests are laboratories. Farmers and healers are researchers. "Every season is an experiment" (The Crucible Group, 1994). Thus, in order to achieve the stupendous array of plant genetic materials currently at the disposal of modern plant breeding, traditional farmers "employ taxonomic systems, encourage introgression, use selection, make efforts to see that varieties are adopted, multiply seeds, field test, record data and name varieties, and in fact, do what many Northern plant breeders do" Friends of the Earth, 1995).

[16] In January 1975, the Board of the American Association for the Advancement of Science passed the following Resolution:

> [B]e it resolved that the Council of the Association (a) formally recognizes the contributions made by Native Americans in their own traditions of inquiry to the various fields of science, engineering, and medicine, and (b) encourages and supports the growth of natural and social programs in which traditional Native Americans' approaches and contributions to science, engineering, and medicine are the subject of serious study and research (see Landy, 1977).

[17] During the period of colonial imperialism, Western medicine:

> [W]as taken as a prime exemplar of the constructive and beneficial effects of European rule. Thus, western medicine was to the imperial mind, one of its most indisputable claims to legitimacy. Since western medicine was regarded as *prima facie* evidence of the intellectual and cultural superiority of Europeans, the figure of the medicine man or shaman was often viewed as inimical to social and cultural progress. Indeed the pejorative term 'witch doctor' has come to stand for savagery, superstition, irrationality and malevolence (Balick and Cox, 1996)

[18] Morley further observes, "Throughout the vast range of traditional medical systems are many beliefs and practices which contain an element of techno-empirical knowledge (see Morley (1978) and Muchena and Vanek (1995)).

[19] As indicated in the preceding pages, it is ironic that the racist foundation of the colonial project was destroyed when the German Third Reich of Hitler's Nazis broke the unwritten and fundamental taboo of colonialism: racial denigration and colonization of Christian Europe by fellow Christian Europeans. Having crossed this rubicon, it became clear that the subjugation, colonization and extermination of peoples on the basis of race or presumed cultural inferiority is a gross evil, unacceptable to the international legal order and human morality. It is significant that Nazi Germany's colonization of Europe and its bizarre categorization of races and ultimately, the genocide of the Jews of Europe fueled the outrage which culminated in the joint statement of August 14, 1941 by President Roosevelt and Prime Minister Churchill affirming the right of all peoples to self-determination. See Atlantic Charter, August 14, 1941. Again, in reflecting the prevalent views that non-Europeans were not part of the concept "peoples" deserving of the right of self-determination as articulated under the Atlantic Charter, Prime Minister Churchill protested that as the Prime Minister, he was intent to hold onto the colonies. In his view, he had not "become the King's First Minister to preside over the liquidation of the British Empire." See Speech of Sir Winston Churchill delivered at Mansion House in London on November 10, 1942.

[20] The critical elements of the CBD on protection of indigenous and traditional knowledge of the uses of plants have been replicated in various continental regimes on plant genetic resources (see OAU Model Law, 2000).

[21] Agreed Interpretation of the International Undertaking Resolution 4/89 and Farmers' Rights, Resolution 5/89, of Twenty-Fifth Session, 1989.

[22] There are momentous changes in the ferment of proposals toward the promotion and protection of indigenous peoples' knowledge. For a brief summary of some current suggestions, see, Decision VI/10 on Article 8(j) and Related Provisions, CBD Secretariat, SCBD/SEL/HM/FV, 26 June 2002; Decision VI/24 On Access and Benefit-Sharing as Related to Genetic Resources, June 27, 2002, SCBD/SEL/VN.

REFERENCES

Ackerknecht E (1991) Medicine and Ethnology. The Johns Hopkins Press, Baltimore, MD
Adair J (1997) The bioprospecting question: should the United States charge biotechnology companies for the commercial use of public wild genetic resources? Ecol Law Q 24(1):131–171
Adams N (1993) Worlds apart: The North-South divide and the international system. Zed Books, London

African Model Legislation for the Protection of the Rights of Local Communities, Framers, and Breeders, and for the Regulation of Biological Resources (2000) OAU Model Law. (Available online: http://www.grain.org/brl_files/oau-model-law-en.pdf. [Last visited June 10, 2006]).

Agrawal A (1999) On power and indigenous knowledge. In: Posey D (ed) Cultural and spiritual values of biodiversity: a complementary contribution to global biodiversity assessment. UNEP, Nairobi

Anghie A (2004) Imperialism, Sovereignty and the Making of International Law, Cambridge, UK: Cambridge University Press.

Balick M, Cox PA (1996) Plants, people and culture-the science of ethnobotany. Freeman and Company, New York

Battiste M, Henderson J (2000) Protecting indigenous knowledge and heritage: a global challenge. Purich Publishing, Saskatoon

Battiste M (ed) (2000) Reclaiming indigenous voices and vision. UBC Press, Vancouver

Bedjaoui M (1985) Poverty of the international order. In: Falk R, Kratochwil F, Mendlovitz S (eds) International law: a contemporary perspective. pp. 152–163, Boulder, CO: Westview Press

Bosselman K (1995) Plants and politics: the international legal regime concerning biotechnology and biodiversity. Colo J Int Environ Law and Policy 7

Christina-Butera A (2001) Assimilation, pluralism, and multiculturalism: the policy of racial/ethnic identity in America. Buffalo Human Rights Journal 7

Cobo M (2000) Study of the problem of discrimination against indigenous populations. U.N. Doc.E/CN.4/Sub.2/1986/7 Add.4, U.N. Sales No. E.86.XIV.3.3

Cohen F (1942) The Spanish origin of Indian rights in the law of the United States. Georgetown Law J 31(1):1–21

Convention on Biological Diversity, entered into force on December 19, 1993, reprinted in (1992) 31 I.L.M. 813

Coombe R (1998) Intellectual property, human rights and sovereignty: new dilemma in international law posed by the recognition of indigenous knowledge and the conservation of biological diversity. Indiana Journal of Global Legal Studies. Ind. J. Global Legal Stud 6(1):59–115

Coombe R (1995) The cultural life of things: anthropological approaches to law and society in conditions of globalization. Am Univ J Int Law Policy 10(2)

Coombe R (2001) The recognition of indigenous peoples' and community traditional knowledge in international law. St. Thomas Law Rev 14(2):275–285.

Davidson B (1967) The African past: chronicles from antiquity to modern times. Grosset & Dunlap

Declaration of Belem (1988) Statement 4

Diop C A (1987) Precolonial black Africa. Lawrence Hill Books

Drahos P (1997) Indigenous knowledge and duties of intellectual property owners. 11 Intell Prop. J. 179

Easlea B (1980) Witch-Hunting, Magic, and the New Philosophy: An Introduction to Debates of the Scientific Revolution, 1450–1750. Haverfield Press, Brighton, Sussex

Edsman C-M (ed) (1962) Studies in shamanism. Almqvist & Boktryckeri, Uppsala

Ezeabasili N (1977) African Science: myth or reality. Vantage Press, New York

Food and Agriculture Organization The IBPGR and Its Policy on *In Situ* Conservation, FAO Doc/AGPG:IBFGRI/83/143

Food and Agriculture Organization (2001) International Treaty on Plant Genetic Resources for Food and Agriculture. concluded on 3 November 2001. Available online: http://www.fao.org/ag/crgrfa/itpgr.htm

Fitzpatrick P, Darian-Smith E (eds) (1999) Laws of the postcolonial. Michigan, Ann Arbor

Four Directions Council (1996) Forests, indigenous peoples and biodiversity: contributions of the four directions council. Submission to the Executive Secretary of the Convention on Biological Diversity by the Four Directions Council, Canada, 15th January 1996. Reproduced in, UNEP/CBD/SBTTA/2/7, 10 August 1996

Fowler C, Mooney P (1990) Shattering: Food, Politics, and The Loss of Genetic Diversity. Tucson: The University of Arizona Press

Fowler C (1994) Unnatural selection: technology, politics, and plant evolution. Gorden and Breach, Langhorne, PA, USA

Friends of The Earth (1995) Intellectual Property Rights and The Biodiversity Convention: The Impact of GATT. Berdfordshire

General Act of The Conference of Berlin (1973) Reprinted in Garvin RJ, Betley JA (eds) The scramble for Africa: documents on the Berlin West African conference and related subjects 1884–1885. Ibadan University Press, Ibadan, Nigeria, p 288

Halewood M (1999) Indigenous and local knowledge in international law: a preface to sui generis intellectual property protection. McGill Law J 44(4):953–996

Hall WE (1977) A Treatise on International Law. Butterworths, London

Hardy C (1994) Patent protection and raw materials: the convention on biological diversity and its implications for U.S. policy on the development and commercialization of biotechnology. University of Pennsylvania Journal of International Business Law 15(2):299–326

Hunn E (1993) What is traditional knowledge? In: Williams N, Barnes G (eds) Traditional Ecological knowledge: wisdom for sustainable development. Australian National University, Canberra

Indigenous and Tribal Peoples available online <www.ecocouncil.ac.cr/indig>

International Labour Organization (No. 169) (1989) Concerning indigenous and tribal peoples in independent countries, concluded at Geneva, 27 June 1989, reprinted in 28 I.L.M. 1382. This instrument has not yet come into effect

Isaac E (1970) Geography of domestication. Prentice-Hall Englewood Chiffs, New Jersey

Javis M (ed) (1991) The influence of religion on the development of international law. Martinus Nijhoff The Hague

Jessup P (1973) Non-universal international law. Columbia Journal of Transnational Law 12(3):415–429.

Johannes RE (ed) (1989) Traditional ecological knowledge: a collection of essays. Gland, IUCN

Johnson M (ed) (1992) Lore: Capturing Traditional Environmental Knowledge. Dene Cultural Institute, International Development Research Center, Ottwa

Kameri-Mbote, AP, Cullet P (1999) Agro-biodiversity and international law: a conceptual framework. Journal of Environmental Law 11(2):257–279

Kloppenburg J. (1988) First The Seed–The Political Economy of Plant Biotechnology, 1492–2000. Cambridge University Press, Cambridge

Lagos Observer (1970) February 19, 1885, quoted in Umozurike UO, International law and colonialism. East African Law Review 3

Lama A (2000, January 12) Law to protect native intellectual property. IPS News Bulletin,. Available online http://www.ips.org

Landy D (ed) (1977) Culture, disease, and healing-studies in medical anthropology. Macmillan Publishing Co. Inc., New York

Langley W (1981) The third world: towards a definition. Boston Coll Third World Law J 2:1–28

Lesser W (1997) Sustainable use of genetic resources under the convention on biological diversity: exploring access and benefit sharing issues. CAB International, Oxford

Lindley M (1969) The acquisition and government of backward territory in international law: being a treatise on the law and practice relating to colonial expansion. Negro Universities Press, New York

Lommel A (1967) Shamanism -the beginnings of art. Mcgraw-Hill Book Company, Toronto

Loudon JB (ed) (1976) Social anthropology and medicine. Academic Press, London

Margadant GF (1980) Official Mexican attitudes towards the Indians: an historical essay. Tulane Law Rev 54

Marx K (1977) Capital, vol 1. Vintage Books, New York

McClellan T (2001) The role of international law in protecting the traditional knowledge and plant life of indigenous peoples. Wisconsin International Law J 19

Meyer G, Blum K, Cull J (eds) (1981) Folk medicine and herbal healing. Charles Thomas Publisher, Illinois

Mgbeoji I (2001a) Patents and traditional knowledge of the uses of plants: is a communal patent regime part of the solution to the scourge of biopiracy? Indiana Journal of Global Legal Studies 9(1):163–186

Mgbeoji I (2001b) Patents and plant resources-related knowledge: towards a regime of communal patents for plant resources-related knowledge. In: Islam N et al (eds) Environmental law in developing countries: selected issues. IUCN, Cambridge

Mgbeoji I (2003) The juridical origins of the international patent system: towards a historiography of the role of patents in industrialization. Journal of the History of International Law 5(2):403–422

Mgbeoji I (2005) Global Biopiracy: patents, plants and indigenous knowledge. UBC Press, Vancouver

Miller K et al (1991) Deforestation and species loss: responding to the crisis. In: Matthews JT (ed) Preserving the global environment: the challenge of shared leadership. Norton, New York

Morley P (1978) Culture and the cognitive world of traditional medical beliefs: some preliminary considerations. In: Morley P, Wallis R (eds) Culture and curing-anthropological perspectives on traditional medical beliefs and practices. University of Pennsylvania Press, Pennsylvania

Muchena ON, Vanek E (1995) From ecology through economics to ethno-science: changing perceptions on natural resource management. In: Warren DM, Slikkerveer LJ, Brokensha D (eds) The cultural dimensions of knowledge: indigenous knowledge systems. Intermediate Technology Publications, London

Mutua M (2001) Savages, victims and saviours: the metaphor of human rights. Harvard Inte Law J 42(1):201–245

Mutua M (2000) What is TWAIL? American Society of International Law Proceedings 94:31–39

Nunes K (1995) We can do.... better: rights of singular peoples and the United Nations declaration on the 'rights of indigenous peoples.' St Thomas Law Rev 7(3):521–555.

Oguamanam C (2004) The protection of traditional knowledge: towards a cross-cultural dialogue. Aust Intellect Prop J 15(1):34–59

Okpewho I (1975) The epic in Africa: toward a poetics of the oral performance. Columbia University Press

Onwuekwe CB (2004) The commons concept and intellectual property rights regime: whither plant genetic resources and traditional knowledge? Pierce Law Review 2(1)

Orr D (1999, July 31) India accuses US of stealing ancient cures. The Times, London

Osava M (1999, August 14) Brazil biodiversity: crackdown on eco-pirates. IPS

Rural Advancement Foundation International (1999) Aussies 'Pirate' Other's Genius? Available online http://www.rafi.org/pr/release11.html

Rural Advancement Foundation International (1995a, September/October) Biopiracy update: a global pandemic. *RAFI Communiqué*

Rural Advancement Foundation International (1995b, September/October) RAFI's List of bio-prospectors and biopirates. *RAFI Communiqué*

Roht-Arriaza N (1996) Of seeds and shamans: the appropriation of the scientific and technical knowledge of indigenous and local communities. Michigan Journal of International Law 17(4):919–965

Sanders D (1983) The re-emergence of indigenous questions in international law. Can Hum Rights Yearb 1:3–30

Santiago C et al (eds) (2004) Accessing biodiversity and sharing the benefits: lessons from implementing the convention on biological diversity. IUCN, Gland

Sarma L (1999) Biopiracy: twentieth century imperialism in the form of international agreements. Temple Int Comp Law J 13(1):107–136

Seidl-Hohenveldern I (1963) Transformation or adoption of international law into municipal law. Int Comp Law Q 12(1):88–124

Semali L, Kincheloe J (eds) (1999) What is indigenous knowledge: voices from the academy. Falmer, London

Shiva V (1988) Staying alive: women, ecology and development. Zed Books Ltd., London

Shiva V (1999) Biopiracy: need to change western IPR systems. The Hindu, Wednesday, July 28

Simmonds N (1979) Principles of crop improvement. Longman, New York

Sinha R (1996) Ethnobotany-the renaissance of traditional herbal medicine. Ina Shree Publishers, Jaipur, India

Slater D (1994, December) Contesting occidental visions of the global: the geopolitics of theory and North-South relations. Beyond Law 4(11):97–118

Stanley A (1993) Mothers and daughter of invention. The Scarecrow Press, Metuchen, NJ

Stenson A, Gray T (1999) The politics of genetic resource control. Macmillan Press, London

Tejera V (1999) Tripping over property rights: is it possible to reconcile the convention on biological diversity with Article 27 of The TRIPs agreement? New Eng Law Rev 33(4):967–987

The Convention Concerning The Protection And Integration Of Indigenous And Other Tribal And Semi-Tribal Populations In Independent Countries (1957) 26 June, reprinted in 328 U.N.T.S. 247

The Crucible Group (1994) People, plants, and patents: the impact of intellectual property on trade, plant, biodiversity, and rural society. IDRC, Ottawa

Tillford D (1998) Saving the blue prints: the international regime for plant resources. Case West Reserve J Int Law 30(2–3):373–446

United Nations Environment Programme, Convention on Biological Diversity, Report of the Intergovernmental Committee 2/13, 1994 par. 16

United Nations Treaty Series 993. 3. (1966)

United Nations Treaty Series 1155 339 Available online: http://untreaty.un.org/English/treaty.asp

Universal Declaration of Human Rights, Resolution (1948) 217 A(III) of December

Vienna Convention on The Law of Treaties. Available online: http://www.un.org/law/ilc/texts/treaties.htm

Villaema L (2000) Indians want patent: Chiefs Prepare International Law Suit Against Scientist Who Registered Indigenous Knowledge. ISTOE Magazine. No. 1581, Sao Paulo, 19 January

Vogel J (1994) Genes for sale: privatization as a conservation policy. Oxford University Press, Oxford United Kingdom

Vogel V (1992) American Indian influence on the American pharmacopoeia. In: Johnson M (ed) Lore: capturing traditional environmental knowledge. Ottawa, Dene Cultural Institute, International Development Research Center.

Wiessner S, Battiste M (2000) The 2000 revision of the united nations draft principles and guidelines on the protection of the heritage of indigenous knowledge. St. Thomas Law Rev 13(1):383–390

Wiessner S (1999, Spring) Rights and status of indigenous peoples: a global perspective and international legal analysis. Harv Hum Rights J 12: 57–128.

Wilkes G (1977) The World's crop germ plasm- an endangered resource. Bull At Sci 8

Williamson E (1992) The penguin history of latin America. Penguin, London

WTO Agreement on Trade Related Aspects of Intellectual Property Rights, 15 April 1994, 33 ILM 1197

PART THREE

IMPLEMENTING ACCESS AND BENEFITS SHARING

CHAPTER 7

LIABILITY PRINCIPLES AND THEIR IMPACT ON ACCESS AND BENEFITS SHARING

INTRODUCTION

The 21st century has been described as the "biotech century" due to advances in health care and agricultural technology (Caulfield and Gold, 2000). These developments have the potential to bring great benefits to society, although they are often accompanied by poorly-understood risks. In turn, harms produced by these technologies can expose developers of modern technology to novel types of liability. For example, in the fields of genetic testing and agricultural technology, patent holders may face potential liability for harm caused by their inventions. A question common to these two sectors is whether patent holders have a duty to inform users of the risks related to their technology and of possible precautionary measures to guard against these risks.

Liability issues related to marketing and use of biotechnology have a major impact on access to, and sharing of, the technology. How the law and market handles risk and liability affects the nature of any access or benefits sharing regimes. Indeed, the scale, scope and distribution of any risks and the extent of resulting liabilities that might arise from the intended or unintended use of a new technology are critical factors taken into account by the technology holder when making arrangements related to access or benefits sharing. The more stringently private law principles regulate the creation of risks and the sharing of information about these risks, the more limitations are indirectly imposed on technological innovation, including the marketing and use of the technology. At the extreme, some may not have the wherewithal to manage risks to the law's satisfaction. Thus, severe liability principles may have an impact similar to that of government regulators closely guarding access to a technology: they can both reduce access to and the scale of net benefits available for sharing.

This chapter addresses whether patent holders of genetic technology in the field of diagnostic medicine and agriculture may indeed be held to a duty to inform under Canadian common law which could force them to compensate injury suffered by users or third parties as a result of its breach. How countries like Canada are likely to address this issue could inspire other systems around the world, especially those whose civil liability rules are similarly governed by common law principles. More precisely, given the shared principles of Canadian, English and American common law Tort principles, there is a real possibility that any principles or precedents laid out in the courts in Canada may have an influence across its jurisdictional frontiers, subject to specific legislation that may exist on product liability.

LIABILITY EMERGING FROM NEW TECHNOLOGIES

Biotechnologies have delivered a series of products and services that are currently used in both the medical and agrifood systems. Two specific cases have highlighted concerns related to the duty to inform: use of genetic testing for various forms of cancer; and the introduction of transgenic food crops in undifferentiated markets.

Genetic Testing

Comprehensive information is crucial to the proper use of genetic testing, as the example of testing for breast and ovarian cancer demonstrates. A significant proportion of victims of hereditary breast and ovarian cancers have mutations of *BRCA 1 or 2* genes. The patents for these genes, as well as for the *BRCAnalysis®* test that identifies genetic alterations to these genes, are held by a single company, Myriad Genetics. While only a small percentage of women who have these mutations will actually develop cancer, women found to be carrying them may choose to lower their risk of disease by: having their breasts or ovaries removed (prophylactic mastectomy or oophorectomy); taking tamoxifen, an anti-oestrogen anticancer agent; taking birth control pills that can reduce risk of ovarian cancer; or participating in extensive mammography and screening programs. Education and counselling of women is therefore crucial given the limited options available and the invasiveness of most of them. Moreover, given the specialised nature of knowledge about genetic testing, women are greatly dependent on the information provided by the patent holder and vulnerable to marketing techniques. Treating physicians are also in a vulnerable position as they rely primarily on the patent holder for their information about genetic tests.

The information necessary for women to make informed decisions, and for physicians to provide adequate counselling, includes the accuracy of the test results, the involvement of genetic mutation on the risk of developing breast and ovarian cancer, and the meaning of a positive or negative result for a particular individual. The importance of knowledge transmission is increased by the adverse impact the test may have on women, including: the psychological effect of learning that they have a higher than average risk of cancer; the temptation, if tested positive, to get mastectomy or oophorectomy although they have only a chance of developing cancer; and the assumption, if tested negative, that they will never develop cancer, which may lead them to neglect other diagnostic techniques.

Escape of Genetically Modified Organisms (GMO)

The area planted in Genetically Modified (GM) crops has increased dramatically since the mid-nineties, provoking an explosion of concern over potential health and environmental impacts (Peterson et al., 2000; Repp, 2000). Although there are divergent opinions on the benefits and risks of GM crops, the possibility of accidental mixing of GM with non-GM organisms is a widely-acknowledged risk that has already been realised on a number of occasions. Such "genetic drift" (Repp, 2000) can occur if GM seeds are inadvertently sown in a non-GM field through wind or insects, or if crops commingle in storage and transport after harvest. For non-GM farmers, crop contamination can lead to: decontamination costs for soil, farming, storage and harvest equipment, transportation facilities, grain elevators and seed supplies; costs for testing and segregating their crops; loss of the ability to farm organically; and, loss of the opportunity to participate in certified organic markets. Precautions that may avoid these losses include: providing accurate information to GM farmers about the risk of cross-pollination or commingling and of resulting injury to non-GM farmers; effective containment measures, such as the necessity of establishing adequate buffer zones and other segregation methods; and better grain-handling systems.

Both patients and non-GM farmers suffering the types of losses outlined above may seek compensation from the patent holders of the genetic technology. In doing so, their strongest juridical argument may be the patent holder's omission to provide the user of the technology with accurate and complete information necessary to avoid the occurrence of injury (Davies and Levine, 2000).

EXISTING CASE LAW ON THE LIABILITY OF PATENT HOLDERS

Private law rules complement our regulatory systems by imposing boundaries on risk creation and granting compensation for injuries created as a result of risk-taking. The two main legal traditions of today's world, civil law and common law, encompass virtually all of the countries that now use or are likely to use new biotechnologies. As several civil law jurisdictions (e.g., in the EU and the Canadian province of Québec) have addressed liability issues related to products and new technologies through legislation, discussion as to the role of the general civil liability principles in regulating conduct in this area have tended to be concentrated in common law jurisdictions, such as Canada and the US, where, incidentally, adoption of these technologies has been more extensive. The courts in those two common law countries are currently grappling with the challenge.

So far, however, there is little Canadian and American case law on the liability of patent holders of genetic biotechnology and, *a fortiori*, on their potential responsibility for breach of a duty to inform. In the area of agricultural genetics, an American class action lawsuit against Aventis for damages claimed to have resulted from the presence of genetically modified "StarLink" corn in the food chain was settled out of court in April 2003 for US$110 million after a motion to dismiss claims based on negligence, public nuisance, and private nuisance was rejected (*In Re StarLink Products Liability Litigation*, 2002). In May 2005, an application to have a claim for compensation of financial injury caused by genetic drift certified as a class action was denied in the Canadian case of *Hoffman & Beaudoin* v. *Aventis and Monsanto*. In this suit, all certified organic grain farmers of Saskatchewan had alleged the loss of their ability to grow and market organic canola due to the contamination of their crops caused by cross-pollination from open-pollinated varieties of GM canola. The claim was partially based on the lack of patent holder's warnings to GM farmers of: the potential harm to neighbouring crops caused by

GM volunteer canola; the necessity of buffer zones to minimise the flow of pollen to surrounding crops; the necessity of ensuring that all farm trucks transporting the seed are properly and securely tarped; the importance of thoroughly cleaning all farm machinery before leaving a field where the GM crop was being grown; as well as of warning neighbours that GM volunteers might emanate from the GM crop.

Smith J of the Saskatchewan Court of Queen's Bench dismissed the application on the basis that it failed to prove that all organic farmers have been financially hurt by genetically modified canola since, 10 years after the introduction of genetically modified canola, some organic farmers are still growing canola and finding markets for it. In her opinion, there was on the facts no evidence that a majority or even a significant minority of the proposed class of organic farmers had suffered loss due to the inability to produce canola sufficiently free from GMO contamination to be marketed as organic. Moreover, while some farmers may have been hurt, she thought there was no evidence that organic farmers as a class have suffered and, consequently, the class action procedure was not justified. The Court also believed that the claim did not disclose a plausible legal basis for imposing liability on the defendant on the grounds of negligence (for want of duty of care), nuisance, *Rylands* v. *Fletcher* and trespass. Although foreseeability of the plaintiff's organic farmers was demonstrated, there was not a sufficiently proximate relationship between the manufacturers and the organic farmers to warrant a duty of care in favour of the organic farmers on the part of the manufacturers. Moreover, the plaintiffs in that case did not attempt to link their financial loss to physical injury caused to their crop, thus claiming essentially compensation for pure economic loss which, if allowed, could, in the opinion of the court, expose the defendants to indeterminate liability (*Hoffman* v. *Monsanto*, 2005, par. 77).

The case law on patent holders' liability for genetic testing is also scarce, although parallels drawn with cases involving other types of diagnostic tests are informative. For instance, in the US case *Rufer* v. *Abbott Labs* (2003), a misdiagnosis led Mrs. Rufer to receive unnecessary chemotherapy, a hysterectomy, and the removal of a portion of her lung. In response to a product liability action, Abbott Labs ("Abbott") and the medical centre where she received her treatment were ordered to pay US$16.2 million in damages to the Rufers on the basis, *inter alia*, that Abbott had breached its obligation to warn by failing to provide adequate information about the potential for false positive results from its test, which was used to make the diagnosis. Although Abbott had reports of several complaints concerning this risk, it was not known by the medical community. The package insert warnings were not sufficiently specific to alert treating physicians to the risk of false positives.

These isolated cases unfortunately provide little guidance. However, the principles of the Canadian tort of negligence, particularly those related to product liability, provide a more solid legal foundation for imposing a duty to inform on the patent holder's shoulders.

PRODUCTS LIABILITY RULES

General Principles

The legal field of Canadian product liability, especially in relation to medical and pharmaceutical products, provides a series of helpful principles that may be applicable to the elaboration of a biotechnology patent holder's duty to inform. Where the patent holder is also the manufacturer of the technology involved in the production of the injury, these principles may find direct application.

In order for the manufacturer to have a duty to inform, there must be, in addition to the injury, a duty of care owed by the manufacturer to the patient/non-GM farmer and a corresponding breach of this duty of care causing the injury. Canadian common law allows for a duty of care to arise where the plaintiff was a reasonably foreseeable victim of the defendant's activity and if there are no policy considerations against imposing liability (*Anns* v. *Merton*, 1978; *Kamloops* v. *Nielsen*, 1984; *Cooper* v. *Hobart*, 2001). The plaintiff, as a particular individual, needs not to have been contemplated. It is sufficient that the activity create a foreseeable risk of injury to persons in the plaintiff's general class.

Canadian law recognises that a manufacturer's duty of care towards his consumers includes a duty to warn of the risks associated with the use of its product of which it has knowledge or ought to have knowledge (*Lambert* v. *Lastoplex*, 1972; *Buchan* v. *Ortho*, 1986; Klar, 2003). A general warning is not sufficient and the duty is ongoing; it also can be triggered by information that becomes known after the product has been in use (*Hollis* v. *Dow Corning*, 1995). This duty extends not only to dangers that are inherent in the ordinary, intended use of the product, but also to risks that flow from the foreseeable use of the product, even if it was not intended by the manufacturer (Klar, 2003). The respect of rules, statutes and regulations about warnings to be given does not necessarily discharge the manufacturer's duty to warn (*Buchan* v. *Ortho*, 1986). Finally, the duty may extend to include third parties who are injured by the manufacturer's failure to warn the immediate user of the product, as long as they were reasonably foreseeable victims (Generally: *Donoghue* v. *Stevenson*, 1932).

Once a duty to warn is recognised, the warning must be adequate. It should be communicated clearly and understandably in a manner calculated to inform the user of the nature of the risk and the extent of the danger. The extent of the disclosure should be commensurate with the potential danger. Finally, it should not be neutralised or negated by collateral efforts on the part of the manufacturer, such as through marketing techniques for instance (*Buchan* v. *Ortho*, 1986; *Hollis* v. *Dow Corning*, 1995).

The manufacturer's knowledge or deemed knowledge, and the correlative absence of the consumer's and doctor's knowledge, is the fundamental basis of the medical manufacturer's duty to warn (*Hollis* v. *Dow Corning*, 1995). In the specific field of pharmaceutical products, the justification for this duty is articulated as the following (*Buchan* v. *Ortho*, 1986): (i) the manufacturer *occupies the position of an expert* in the field; (ii) this requires that the manufacturer be under a *continuing duty to keep abreast of scientific developments* pertaining to its product through research, adverse reaction reports, scientific literature and other available methods and to make all reasonable efforts to communicate the additional information to prescribing physicians; and (iii) unless doctors have current, accurate and complete information about a drug's risks, their ability to exercise the fully-informed medical judgment necessary for the proper performance of their vital role in prescribing drugs for patients may be reduced or impaired. This information imbalance also exists between the pharmaceutical manufacturer and physicians (and, by extension, other intermediaries) because of their limited pharmaceutical knowledge (Peppin, 1991).

Proving a breach of the duty to warn requires that it was reasonably foreseeable that injury could result from the withholding of information and that there existed reasonable precautions against this occurrence. The combination of the seriousness of the potential injury with the straightforward nature of the reasonable precautions necessary to avoid it weighs in favour of finding breach of the duty on the part of the patent holder (*Bolton v. Stone*, 1951; *Wagon Mound 2*, 1967).

Application to Genetic Testing and GMO Escape

Knowledge imbalance and dependency on the manufacturer's information are thus the key elements in the imposition of a duty to warn under medical and pharmaceutical product liability principles. These factors are also found in the genetic testing and GMO escape hypotheses. Moreover, in both circumstances, injury to the victim is clearly foreseeable. In addition, the class of people likely to be affected is well-defined, namely patients using genetic tests, and organic and conventional farmers (*In Re StarLink Products Liability Litigation*, 2002). Policy considerations supporting a finding of liability include the fact that manufacturers, having full knowledge of the possible risks, are best able to structure insurance to guard against them (Endres, 2000) and that the imposition of a duty of care on the shoulders of biotechnology companies could encourage them to develop and take additional precautionary measures before and after making their technology available to the public. However, other policy factors may militate against a finding of duty of care, including the desire to shield the biotechnology industry from overburdening liability (Burk and Boczar, 1994). As for breach of the duty, the potential injury to the plaintiff is serious in both cases: possibly a mastectomy or the destruction of an entire crop (Endres, 2000). Moreover, the cost of providing precautionary information is relatively limited. A patent-holder/manufacturer must simply supply a clear and forthright warning of the dangers inherent in the use of their products of which it knows or ought to know. Consequently, there are ample justifications for the parallel drawn between medical/pharmaceutical products liability and the imposition of a duty to inform in genetic testing and GMO escape cases.

In the case of GM escape, however, there are also some important distinctions to be made. The possible risks discussed here are of property damage and financial loss, but not injury to one's physical integrity, which constitutes an important justification for imposing a stringent duty to warn on medical products manufacturers.

In addition, the chain of transfer of technology is more complex and indirect than the typical manufacturer/doctor/patient relationship and may involve several other stakeholders such as seed breeders or distributors. One may thus ask whether transmission of information to the "middle man" should be deemed sufficient, thereby transferring the duty to inform onto the intermediary's shoulders. These two issues will be discussed below. Finally, the risk of injury in the GMO escape cases is to a third party who did not intend to use or benefit from the product, and who consequently does not have the choice to undertake or refuse the risk. However, this fact should strongly support the imposition of a duty to inform given the non-consensual nature of the risk.

Thus, given the inequality of information between users and manufacturers, the foreseeability of the injury to third parties that can be caused by lack of information, the well-defined class of people likely to be affected, and the low burden imposed by the duty to share information, general negligence principles should ground the imposition on biotechnology companies of a duty to exercise reasonable care by educating GM farmers about the best risk management practices. This obligation may even go as far as requiring the obligation to see to the implementation of these practices through contractual arrangements with the GM farmers (Kershen, 2004). This opinion may be challenged by the common law's reluctance to impose duties to control the conduct of third parties outside of a special relationship. Against all of the claims above, there are counterarguments that must be addressed.

Counterarguments

There are several counterarguments in favour of the patent holder, but only three will be briefly discussed below. They relate to: (i) the learned intermediary doctrine; (ii) causal difficulties; and (iii) pure economic loss.

i. *Learned intermediary doctrine*
Where the product is a highly technical one that is intended to be used only under the supervision of an expert, a warning to the expert (the "learned intermediary") will suffice (*Buchan* v. *Ortho*, 1986; *Hollis* v. *Dow Corning*, 1995). This rule, however, presumes that the intermediary is "learned", i.e. that the learned intermediary's knowledge *approximates that of the manufacturer* (*Hollis* v. *Dow Corning*, 1995). In genetic testing cases, this rule may apply where the doctor prescribing the test and counselling the patient has the same knowledge as the developer of the technology. Although it was originally developed to apply to pharmaceutical manufacturers, the transposition of this rule to GM escape is also possible. The "learned intermediary rule", in this context, may simply constitute a specific application of the *novus actus interveniens* maxim; an independent negligent action from a third party that has a causal relation to the plaintiff's injury "may" in some cases "break" the chain of causation between the original wrongdoer and the victim. Thus, the question becomes whether the expertise of a specific GM farmer may be sufficient to argue that his lack of preventive measures against cross-pollination, in full knowledge of the risk and the precautions to reasonably guard against it, is sufficient to break the chain of causation.

ii. *Causal difficulties*

The causal demonstration in both hypotheses may be complicated by the fact that causal connection depends on: (i) the hypothetical behaviour of the plaintiff (in genetic testing cases only); (ii) the hypothetical behaviour of a third party; and (iii) the physical origin of the contamination (in GMO escape cases only). According to traditional causation principles, it is for the plaintiff to argue that if the information had been provided, she would have acted in a way that would have prevented the injury from occurring, or would have reduced its extent. For instance, the patient must demonstrate that, had she been informed of the risks related to genetic testing, she would not have undertaken the test or, in light of a positive result, submitted to the mastectomy. In order to assist plaintiffs in this demonstration, the Supreme Court of Canada has departed, in a pharmaceutical product liability case, from the causal approach usually applied in medical informed-consent cases, where one must demonstrate what the reasonable patient placed in the same circumstances would have done (*Reibl* v. *Hughes*, 1980; *Hopp* v. *Lepp*, 1980). It rather applies a subjective test and asks what this *specific* patient would have done if information had been provided. This pro-plaintiff causation test is justified by the fact that manufacturers can be expected to act in a more self-interested manner than doctors and, consequently, to overemphasise the value of their product and underemphasise the risk (*Hollis* v. *Dow Corning*, 1995).

In both genetic testing and GMO escape cases, assessing the causal relationship between the fault and the injury necessitates proof of the hypothetical behaviour of a third party. In the GMO escape hypothesis, it requires the demonstration that if he had been provided with the relevant information by the patent holder, the GM farmer would have diligently followed all of these recommendations resulting in the total or partial avoidance of the injury. In the genetic testing hypothesis, the third party is the attending physician who may, for some reason, omit to disclose the information provided by the patent holder. Demonstrating what a third party would have done had the manufacturer disclosed the information was deemed too onerous a burden for plaintiffs in *Hollis* v. *Dow Corning* (1995). The burden of proving the hypothetical behaviour of the third party was therefore reversed onto the manufacturer's shoulders, a solution that may equally apply here.

In the GMO escape hypothesis, it may often be impossible to determine where and how the escape or mixing occurred, thereby preventing the injured farmer from identifying the chain of factual causation leading to the responsible patent holder. This may have little bearing on the patent holder's responsibility as long as it is possible to link the exact GM crop to a specific patent (through genetic markers, for instance). Even if the manufacturers are identifiable, there may be multiple wrongdoers given that several possible pollination sources may exist that are linked to different manufacturers/patent holders, in addition to the GM farmer's possible faulty contribution. This difficulty may be easily resolved through the application of the "material contribution test"; that is, as long as the patent holder has contributed in a more than minimal way to the contamination, he can be liable for the whole injury (*Athey* v. *Leonati*, 1996).

iii. *Pure economic loss*

A last problem lies in the general reluctance of common law courts to allow claims for pure economic loss. These cases arise when individuals who have suffered neither personal injury nor property damage assert that another's negligence has resulted in their financial detriment. Since loss of profits is the typical head of damage in the GMO escape hypothesis, qualification of the loss as purely economic could lead to the straightforward rejection of the claim (e.g. *Sample v. Monsanto Co.* [2003], where no evidence of physical damage to the crop had been provided). Some commentators argue that conventional or organic farmers are unlikely to be able to point to any property damages in most cases (Lee and Burrell, 2002). However, if a plaintiff can link his financial injury to physical injury to his crop on the basis that cross-pollination physically alters the substance of a plant or the content of a harvest, the financial nature of the damage will not be an obstacle to recovery. For instance, in *Starlink Corn Products Liability Litigation*, 2002, the court concluded that there is a viable claim for physical harm to the crops where they are contaminated by pollen from GM crop on a neighbouring farm or where the non-GM farmer's harvest is contaminated by commingling with GM crops in a transport or storage facility. Moreover, the impossibility in the GMO cases to attribute responsibility to third parties through contract law, and the absence of a threat of floodgates, which are typical policy arguments justifying exclusion of liability for pure financial losses, may weigh in favour of allowing recovery despite the qualification of pure economic loss, given the more generous Canadian approach to these types of losses (Glenn, 2003).

CONCLUSION

Canadian tort law appears flexible enough to accommodate new problems created by emerging technological and scientific developments and to contribute to the taking of greater precautions and assumption of greater social responsibility and accountability by patent holders. Such precautions lie, amongst others, in the effective transmission of information by the patent holder to avoid potential harm to users of the technology (patients) or third parties (organic or conventional farmers). On the basis of parallels drawn with the area of medical and pharmaceutical product liability, patent holders are likely to be held to a stringent standard in terms of duty to inform of risks and precautionary measures against harm caused by their technology to doctors and patients in the context of *genetic testing*, and to GM farmers in the context of *GMO escape*. More precisely, it should be the patent holder's duty to adequately inform the user of their technology (GM farmer, patient, doctor) of the risks associated with its use and to take the appropriate precautions against realisation of these risks of which the patent holder has knowledge or ought to have knowledge. A general warning should not be sufficient and the duty to inform should not be neutralised or negated by collateral efforts on the part of the manufacturer. This duty should also include the obligation to keep abreast of scientific developments related to these risks, which implies a duty to play a responsible role in identifying these risks through adequate research and testing.

This position may find its justification in the fact that: (i) patent holders/manufacturers, in both cases, control the information on the risks of injury and precautions needed to avoid it; (ii) the user of the technology (GM farmer, patient or doctor) is, in most cases, at a disadvantage in terms of knowledge about the technology, the latter depending on the information provided by the patent holder; (iii) the patent holder controls, or can control, the conditions in which the activity is exercised; (iv) the patent holder creates the risk; (v) for commercial profit; and, finally, (vi) the patent holder is the party with the best ability to avoid the loss, and to spread it if it occurs.

Courts should recognise that patent holders share responsibility for the taking of precautions against injury flowing from the marketing of their technology, if only on the basis of their knowledge and the empowerment that this knowledge gives them, as well as on the basis of the important financial benefits they obtain from risk creation and limitation of the available information to users. Reflection on liability issues related to this field may play its part in the adoption of more responsible corporate practice on the part of biotechnology companies in terms of knowledge sharing and the taking of precautionary measures.

How these private law systems handle the questions of assignment of liability is likely to have a major impact on the trajectory of these new technologies, ultimately determining how, where, when the technologies will be developed and commercialised and who will ultimately benefit from their use. One must consequently be conscious of the fact the imposition of rigid liability rules on the technology holder and user may have a proportional impact on the extent of the limitations imposed on the marketing, use, and sharing of the technology, and more generally on technological innovation. Any adoption of the principles developed above must thus take into account their potential negative impact on access to, and benefit sharing of, genetic testing technology and transgenic food crops.

REFERENCES

Burk DL, Boczar BA (1994) Symposium: biotechnology and tort liability: a strategic industry at risk. Univ Pittsbg Law Rev 55(3):791–864

Caulfield TA, Gold ER (2000) Whistling in the wind. Forum Appl Res Public Policy 15(1):75–79

Davies JA, Levine LC (2000) Symposium: biotechnology and the law: biotechnology's challenge to the law of torts. McGeorge Law Rev 32: 221

Endres AB (2000) "GMO": genetically modified organism or gigantic monetary obligation? The liability schemes for GMO damage in the united states and the European union. Loyola LA International & Comparative Law Review Loy. L.A. Int'l & Comp. L. Rev 22(4):453–505

Glenn JM (2003) Genetically modified crops in Canada: rights and wrongs. Journal of Environmental Law and Practice J Environ Law Pract 12:281

Kershen DL (2004) Legal liability issues in agricultural biotechnology. Crop Sci 44(2): 456–463

Klar L (2003) Tort Law. 3rd edn. Thompson, Toronto

Lee M, Burrell R (2002) Liability for the escape of GM seeds: pursuing the 'victim'? Mod Law Rev 65(4):517–537

Peppin P (1991) Drug/vaccine risks: patient decision-making and harm reduction in the pharmaceutical company duty to warn action. Can Bar Rev 70(3): 473

Peterson G et al (2000) The risks and benefits of genetically modified crops: a multidisciplinary perspective. Conserv Ecol 4(1):13

Repp RA (2000) Biotech pollution: assessing liability for genetically modified crop production and genetic drift. Ida Law Rev 36(3):585–620

CASES

Anns v. Merton (1978) A.C. 728 (H.L.)
Athey v. Leonati (1996) 3 S.C.R. 458
Bolton v. Stone (1951) A.C. 850 (H.L.)
Buchan v. Ortho Pharmaceutical (Can.) Ltd. (1986) 35 C.C.L.T. 1 (Ont. C.A.) at para. 17–18, 21, 27–28, 35, 52–53, 63
Cooper v. Hobart (2001) 3 S.C.R. 537
Donoghue v. Stevenson (1932) A.C. 562 (H.L.)
Hoffman v. Monsanto Canada Inc. (2003) Sask. Q.B. April 10, 2003, 2003 Sask. D.J. Lexis 130
Hoffman v. Monsanto Canada Inc. (2005) S.K.Q.B. 225
Hollis v. Dow Corning (1995) 4 S.C.R. 634 at para. 23, 25–26, 29, 45–46, 55
Hopp v. Lepp (1980) 2 S.C.R. 192
Kamloops v. Nielsen (1984) 2 S.C.R. 2
Lambert v. Lastoplex Chemicals co. (1972) S.C.R. 569
Overseas Tankship v. Miller Steamship (Wagon Mound 2) (1967) 1 A.C. 617 (P.C.)
Reibl v. Hughes (1980) 114 D.L.R. (3d) 1 (S.C.C.)
Rufer v. Abbott (2003) 118 Wn. App. 1080, 2003 Wash. App. LEXIS 3256
Sample v. Monsanto Co. 283 F.Supp.2d 1088 (ED Mo 2003) at 1093
Smith v. Leurs (1945) 70 C.L.R. 256 at 262 (Aust. H.C.) 2003
StarLink Corn Products Liability Litigation 2002 (Kramer v. Aventis CropScience USA Holding Inc. 212 F.Supp.2nd 828 (N.D. Ill. 2002) at 842

CHAPTER 8

BEYOND THE RHETORIC: POPULATION GENETICS AND BENEFIT-SHARING

INTRODUCTION

Information derived from the Human Genome Project (HGP) promises to inspire an array of future innovations with profound societal benefits (Collins et al., 2003). Population genetic research is now viewed as a necessary next step in the evolution of research based on the human genome (Mathew, 2001; Kaiser, 2002; Chakravarti, 1999). It is anticipated that projects will be developed in which the DNA from thousands of research subjects will be linked to medical records and to genealogical data; these complex databases will provide much needed insight into the etiology and prevention of many complex human diseases. Analyses of data procured in large-scale population genetic studies will enable researchers to gain a better understanding of the gene-environment interactions that are now implicated in cardiovascular diseases, metabolic disorders, musculoskeletal diseases, neuropsychiatric diseases and cancer. The completion of the sequencing phase of the HGP provides researchers:

> with an unparalleled opportunity to advance our understanding of the role of genetic factors in human health and disease, to allow more precise definition of the non-genetic factors involved, and to apply this insight rapidly into the prevention, diagnosis and treatment of disease.... [T]he time is right to develop and apply large-scale genomic strategies to empower improvements in human health, while anticipating and avoiding potential harm.[1]

The challenge remains to justly translate the potential of the HGP into improved human health and well-being.[2]

To facilitate the understanding of gene-environment interactions, numerous large-scale population genetic research studies have been commenced or are in the process of being developed.[3] Prospectively gathered and appropriately stored biological materials coupled with comprehensive measures of environmental exposure, phenotype and genotype will help researchers to confirm or refute existing assumptions about the interrelatedness of these factors. It is hoped that such new knowledge will facilitate the planning of health promotion and disease prevention at the level of populations as well as render adverse drug reactions predictable and avoidable based on an individual's genotype.[4]

The General Concerns

However, with this optimism about the long-term potential of population genetic research come numerous ethical, legal and social issues; concerns regarding individual and group consent (Beskow et al., 2001; Caulfield, 2002; Deschenes et al., 2001; Weijer, 1999, 2000), ownership of human biologic materials (Gold, 1996; Harrison, 2002; Litman & Robertson, 1996), privacy and confidentiality (Annas, 1993; Gostin & Hodge, 1999; Human Genetics Commission, 2002; Robertson, 1999), genetic discrimination and stigmatization (Greely, 1997; Juengst, 1988; Markel, 1992) and eugenics (Anonymous, 1995; Wertz, 2002; Wertz & Fletcher, 1998) have been repeatedly raised. Academic researchers focussing on the "ethical, legal and social issues" (ELSI) of the HGP have initiated much debate on these topics. It has been noted that failure to apply the highest ethical standards to population genetic studies could undermine public trust and confidence in scientific development and the products of medical research (Blumenthal, 1996; Lowrance, 2001). Moreover, the increased involvement of the private sector in human genetic and genomic research may promote secrecy amongst researchers, inhibit academic freedom, diminish the collaborative will of researchers and inappropriately shift the focus of research from basic discovery to commercializable end products (Cho et al., 2003; Eisenberg, 1992).

The Particular Problem of Property

Concepts of ownership, values, rights and sharing of knowledge vary from one culture to another, between political systems and between legal systems. These differences have particular relevance to the collection and banking of human biological samples: should they be characterized "person", "property", "*sui-generis*"—or not at all; should human genes be patentable; what is the legal status of linked and linkable databases? These questions have been imperfectly addressed. For example, no consensus exists as to who owns donated DNA samples. In the present age of globalization, it is important that these fundamental conceptual differences are not ignored.

Views on intellectual property protection differ substantially between the North and the South. The South tends to view the accumulation of intellectual property on human genetic material as antithetical to their world-view wherein "[t]he idea of a better ordered world is one in which medical discoveries will be free of patents and there will be no profiteering from life and death" (Gandi, 1988). As one commentator has noted, "[t]he forms and very definition of ownership [in the South] are crafted in a way opposite to conceptions of western legal and economic structure central to the development of private and public law" (Gana, 1995). The industrialized North tends to favour strong intellectual property protection and a broad interpretation of patentable subject matter.[5] Critics of the Northern world-view argue that Western powers are driven by the impulse to "discover, conquer, own, and possess everything, every society, every culture" and that "the colonies have been extended to the interior spaces, the 'genetic codes' of life-forms from microbes and plants to animals, including humans" (Hamilton, 1997; Shiva, 1997, 1999). However, the fact that some "Southern" states, including Brazil have utilized intellectual property to protect non-human biological materials bears comment (Pennisi, 1998). Indeed, this perceived dichotomy between the North and South may be over-exaggerated as demonstrated by the more nuanced position adopted by the World Health Organization with respect to "Genomics and World Health" (WHO, 2002).

Heated debate has ensued over the double standard that permits researchers and biotechnology companies to profit from scientific discoveries that rely on the gratuitous donation of human tissue for research. This issue has been discussed in relation to the Moore case (1988) and more recently the issue has arisen in the Canavan case (2002).[6] Many people question the ethical appropriateness and legal defensibility of strong intellectual property protection for biotechnological inventions. Even among those who view intellectual property protection as an important economic stimulator, there is disagreement as to the appropriate scope and duration of patent protection (Barton, 1997; Merges and Nelson, 1990). Whether patent protection for products of biotechnology is, on the whole, beneficial or detrimental is not fully understood (Barton, 1997; CIPR, 2002). In particular, the effect of strong intellectual property protection, as required by the TRIPs Agreement, on the long-term ability of the developing world to deliver adequate healthcare and provide incentives for research is uncertain (CIPR, 2002).

Neo-Colonialism?

Some suggest that population genetic studies may provide yet another mechanism for the developed world to benefit at the expense of the developing world (WHO, 2002). Genetic knowledge gathering is often disparaged as "neo-colonialism", "biocolonialism" or "biopiracy". Unfortunately, controversies have arisen that give credence to these unattractive labels (Staples, 2000).[7] Yet, others speak to the creativity of the press in making a story (Harvard School of Public Health, 2003; Pomfret and Nelson, 2000).[8]

It has been suggested that "without explicit attention at the international level, the initial technological fruits of genomics are likely to consist primarily of therapeutic and diagnostic applications for conditions affecting large populations in rich countries" (WHO, 2002). International organizations including UNESCO and the Human Genome Organization take the position that the fruits of the HGP should be shared by all (HUGO, 1996, 2000; UNESCO, 1997). To this end, "benefit-sharing" has become a catchphrase in the legal and ethical discourse on human genetics (Knoppers, 1999, 2000).

The next phase of the HGP is a grand project that has the potential to foster equity and improve the health and well-being of all. The thesis of this chapter is that, if appropriately developed and applied to human population genetic research, a rational model of benefit-sharing can provide a mechanism that will enable cooperation between the developed world and the developing world. A relevant benefit-sharing model can be readily developed through the innovative adaptation of the benefit-sharing provisions contained in the Convention on Biological Diversity (CBD) and informed by declaratory statements, ethical guidelines and professional codes of conduct. Such a model can recognize the relative importance of intellectual property protection and of equitable principles demanding that individuals and groups not be unjustly enriched at the expense of others.

To found this argument, we provide:

- A general explanation of the concept of benefit-sharing;
- an overview of some of the key large-scale population genetic initiatives that are relevant to a discussion of benefit-sharing;
- an overview of relevant legal and ethical norms; and
- a brief overview of the specific provisions of the CBD that may be adapted for use with human genetic resources.

THE CONCEPT OF "BENEFIT-SHARING"

Benefit-sharing is a concept originating in the international law arena. It stems from the overarching concept that certain of the earth's resources are the "Common Heritage of Mankind" (Baslar, 1998) and that the benefits and burdens of exploiting and sustaining those resources ought to be universally shared. It is an evolving concept that supports the conservation and sustainable use of those of the world's resources that are a "collective and vital interest of all mankind" (Baslar, 1998; UN Convention on the Law of the Seas, 1982). It "succinctly expresses—with all its merits and limitations—the 'new model' of world community which has gradually emerged since 1945" (Cassese, 1990).

It can be effectively argued that the human genome at the level of the species is, by definition, "a collective and vital interest of mankind" (Knoppers, 1991). It is both a life-giving resource and an information resource reflective of human history. By analogy to other global resources, it is arguable that the human genome must be protected in the interest of future generations.

The CBD recognizes that the conservation of biological diversity is essential to evolution and the maintenance of life sustaining systems of the biosphere and that biological diversity is "the common concern of humankind." Though not directly applicable to human genetic resources, the CBD can inform the development of a rational benefit-sharing regime concerning human genetic material. Albeit from the perspective of state sovereignty, the CBD is relevant to a discussion of human genetic resources because: (1) it places benefit-sharing in the context of genetic resources; (2) it defines and provides examples of benefits and burdens associated with genetic resources; and (3) it provides a rational starting point for the discussion of benefit sharing in the context of human genetic material at both the national and international levels. The fair and equitable sharing of benefits arising from the utilization of genetic resources is an express objective of the CBD (Knoppers, 1991). An overview of the provisions of the CBD that deal with the practical mechanisms to effect benefit-sharing are discussed in the last part of this chapter.

The CBD and the Bonn Guidelines on Access to Genetic Resources and Fair and Equitable Sharing of the Benefits Arising out of their Utilization (2002) clarify the concept of benefit-sharing in the context of non-human genetic resources. Monetary and non-monetary benefits that may be included in benefit-sharing agreements may include payment of access fees, royalties, license fees, research funding, joint ventures, joint ownership of intellectual property rights, sharing of information, research collaboration, contribution in education, technology transfer, capacity building (human resources and institutional), social recognition and joint ownership of intellectual property rights.

The Bonn Guidelines suggest that benefits be shared with "all those who have been identified as having contributed to resources management, scientific and/or commercial process." Governmental, non-governmental or academic institutions and indigenous and local communities may be included in sharing arrangements.

In its 2000 Statement on Benefit-Sharing, the Human Genome Organisation Ethics Committee defines "benefit" as a good that contributes to the well being of an individual and/or a given community (HUGO, 2000). What constitutes a benefit for a particular individual or community will depend on the circumstance, the needs, the values, and the cultural priorities and expectations of that individual or that community. It does not simply define monetary benefit. Some mechanisms that have been envisioned to effect benefit-sharing include:

> agreements with individuals, families, groups, communities or populations that foresee technology transfer, local training, joint ventures, provision of health care or of information infrastructures, reimbursement of costs, or the possible use of a percentage of any royalties for humanitarian purposes (HUGO, 1996).

The Role of Industry

Private industry has become critical to research and development that will translate the information gleaned from the HGP into tangible products and procedures that will benefit individuals and society.[9] However, although private industry contributes proportionally more to genetic research than do governments the reality is that public funding gives industry a jump-start on the commercialization path. Increasingly, publicly funded research will be commercialized by private industry (Caulfield et al., 2003). Because of this, some argue that firms that benefit from publicly funded research have a moral obligation to repay society (Berg, 2001; Weijer, 2000). In particular, where populations or communities have contributed to research projects and profits ultimately accrue to the commercial entity (or entities) involved, it is appropriate to consider whether and how profits (or other benefits) ought to be shared with the broader community.

OVERVIEW OF CURRENT POPULATION GENETIC RESEARCH INITIATIVES

Icelandic Health Sector Database

The most well-known and controversial population genetic initiative is that developed in Iceland. In 1997, Kari Stefansson, the director of deCODE Genetics Inc, proposed to the Icelandic government a method of processing medical data whereby deCODE would create a database utilizing publicly available medical health records of the entire population of Iceland, molecular genetic data, and genealogical data (Greely, 2000). Iceland is ideal for a population genetic study of this nature because of its small population, its genetic homogeneity, its penchant for collecting detailed genealogical data and a political willingness to participate in a project of this nature (Greely, 2000). The Icelandic Ministry of Health viewed the proposal as an opportunity to improve Icelandic health services and passed legislation enabling the creation of such a database (Iceland Act on a Health Sector Database, 1998, and Iceland Act on Biobanks, 2000).[10] The legislation has been harshly criticized because it permits access to publicly created medical health records of all Icelanders without prior informed consent. deCODE has, nevertheless, decided to obtain individual informed consent of all participants for access to health records (deCODE, online). Individuals who wish not to have their personal medical histories included in the databank must opt-out of the project (Iceland Act on a Health Sector Database, 1998). Other aspects of the Icelandic project have also come under harsh criticism.[11]

The Icelandic legislation enabling the establishment of the Health Sector Database establishes a scheme whereby a party may be granted an exclusive operating license to create and operate a health sector database (Berger, 1999). Pursuant to the statute, deCODE was granted an exclusive license that gives it the right to exploit the database for a twelve-year term in exchange for an annual license fee. The legislation does not require deCODE to share with Icelanders any money it makes through public offerings of corporate shares. Under the terms of the agreement, deCODE is granted the authority to sublicense data to others and has, in fact, entered into an exclusive sub-license agreement with Hoffman-La Roche that will give the latter exclusive access to the database to explore the genetic origins of 12 diseases (Berger, 1999).

The sub-license agreement between deCODE and Hoffman-LaRoche promises that Icelanders will be provided, free of charge for the patent term, any products that are developed using data from the Icelandic database (Greely, 2000). It has been argued that promises to provide free drugs and diagnostics that are developed through use of the database during the patent period are potentially empty and that it is impossible to foresee how many drugs, if any, will be developed as a direct result of data derived from the DNA of the Icelandic people (Greely, 2000). There is some concern that the Icelandic population has been exploited by its own government and that it has no genuine appreciation of the magnitude of what it has given away (McInnis, 1999). Professor Henry Greely of Stanford University is critical of the agreement; it is his opinion that:

> deCODE has gained an asset of speculative value—it may ultimately be worth billions, or it may be worth nothing. Currently, the market puts a value on deCODE of about $2 billion. Most of that value must stem from its possession of the license to build and use the database. Iceland will gain a national database of health records, whose value to Iceland, either during the term of the license or afterwards, is speculative, and the possibility of a thriving biotechnology company in Reykjavik. Both sides may do poorly in the deal; both may do well. The questionable aspect of the agreement is that deCODE's potential profit is much greater than Iceland's potential benefit. deCODE and its shareholders, a substantial minority of whom are U.S.-based venture capital firms and non-Icelandic employees and consultants, may make billions. Four hundred jobs may be a fair return to Iceland for a modestly valuable database; it seems inadequate compensation if deCODE achieve their goals (McInnis, 1999).

In addition, the fact that the primary responsibility of the directors of deCODE is to its shareholders and not towards the population of Iceland is ethically, if not legally, problematic. Whether the Icelandic government has struck a fair deal on behalf of its citizens remains to be seen.

Estonian Gene Bank Project

The Estonian Genome Project is a public-private partnership developed by scientists under the direction of the Estonian Genome Foundation—a not-for-profit organization founded in 1999 by Estonian scientists, physicians and politicians to support genetic research and biotechnology in Estonia. The Estonian Genome Foundation aims to establish a national gene bank in Estonia comprised of databased genotype and phenotype data reflective of the Estonian population. Unlike the Icelandic population, the Estonian population is heterogenous and is thought to be an ideal representation of European populations generally (Dawson et al., 2002).[12] As in Iceland, Estonia has a comprehensive national health information infrastructure that will be used to facilitate the Estonian Genome Project. It is hoped that three-quarters of Estonia's population of 1.4 million will participate in the project.

The ethical and legal framework for the EGP is established in the Human Genes Research Act (Estonia HGRA, 2001), passed by the Estonian parliament in 2001. Under the HGRA, voluntary informed consent is required of all participants. The EGP has created an advisory Ethics Committee; the members of this committee are experts in medical ethics and law. The purpose of the Committee is to "assist in ensuring the protection of the health, human dignity, identity, security of the person, privacy and other fundamental rights and freedoms of gene donors" (Estonian Genome Project, Online). The HGRA also provides that the database may only be used for scientific research, medical treatment, public health research and statistical purposes. Only the gene donor and his or her physician will have the right to receive personalized information from the database. Gene donors will receive no remuneration for participation in the study. There is a right to withdraw from the project at any time. The Gene Bank database is not to be taken outside of Estonia.

Due to the magnitude of this endeavour, both public and private input has been sought. To secure funding for the project and to deliver tangible pharmaceutical and health related products, a for-profit US entity, EGeen, was created and granted commercial access to all data emerging from the EGP. In exchange, the Estonian Genome Foundation holds a financial stake in the company. On the scope of research, a spokesperson for EGeen has stated:

> Given the representation of virtually all major common diseases in the Estonian population, EGeen will not focus on just some diseases but is [...] seeking to gain insights into virtually any common disease such as cancer, central nervous system and cardiovascular disease. Having such a heterogenous population representation of all Caucasians is also advantageous in that all findings in the Estonian population will be applicable in other populations and will therefore have commercial significance (Habeck, 2002).

EGeen is the corporate commercial arm of the Estonian Gene Bank Project. It plans to develop important intellectual property in the areas of drug and diagnostic targets, pharmacogenetic profiles and related information technology. All information resulting from the Estonian Gene Bank initiative will be transmitted back to the general practitioners in the hope of providing better health care.

UK BioBank

UK BioBank is a massive undertaking, planned and funded by the Wellcome Trust, the Medical Research Council (MRC) and the Department of Health and anticipated to cost £60 million (BioBank UK, Online). Similar to the projects in Iceland and Estonia, this project aims to study the effect of genes and environment on common complex disorders including diabetes, Alzheimer's disease and early onset heart disease. It is planned to collect some 500,000 samples from male and female participants between the ages of 45 and 60. The health of participants will be monitored over 10–20 years. Information collected and databased by the National Health Service will be used to supplement information collected specifically for the Biobank project (BioBank UK, Online).

It is planned that UK BioBank will be managed by a private not-for-profit company jointly owned by the MRC and the Welcome Trust (House of Commons, Science and Technology Committee online). The company will have the authority to license access to data to academic researchers and pharmaceutical companies.

Unlike Iceland, the United Kingdom sponsors of UK Biobank have agreed that no single company will be granted exclusive access to the data. The precise details of the contractual arrangements between UK Biobank and potential commercial licensees have not been publicly disclosed. The Draft Protocol states that "any use of the material from the study by commercial organizations will be subject to the approval of the Scientific Management Committee and the Overseeing Body and must conform to the relevant ethical and legal requirements" (BioBank UK, online). Participants in UK Biobank will be notified of the likely involvement of commercial entities.

The House of Commons Science and Technology Committee (the "Committee") recently launched a scathing report on the work of the MRC. The report is generally critical of the MRC's administration and its "poor financial management and poor planning." It cites the MRC for committing too many funds over long periods leading to the rejection of many top quality grant proposals. Of the 41 recommendations/conclusions made by the Committee, 12 are in direct reference to UK Biobank. The Committee alleges that the scientific case for UK Biobank was made by the funders to support a "politically driven project." The Committee harshly criticizes the peer review process for UK Biobank that was inconsistent with the peer review process for other MRC funded projects. The Committee raises concern over premature allocation of funds to the project and that a commitment had been made "before the scientific questions over its value and methodology were fully addressed."

The Committee also raises concerns over the consultation process that has been undertaken thus far and questions how the results of consultation were used to develop a strategic plan for UK Biobank. The Committee clearly feels that more consultation is needed and that the consultation effort to date has been a "bolt-on activity to secure widespread support for the project rather than a genuine attempt to build a consensus on the project's aims and methods."

On a positive note, the Committee is satisfied that the MRC appears to be taking "a sensible attitude to industrial involvement in UK Biobank." The Committee notes that whilst all results will be put into the public domain that the involvement of industry is "inevitable and necessary" if new therapies are to arise.

GeneWatch[13] provided the Committee with detailed criticism of the scientific merit of UK Biobank; the main objections were: (1) the size of the cohort—they postulate that a smaller more focussed study would be more useful; (2) the inadequacy of the medical records in providing necessary and relevant information; (3) the advanced age of the cohort would make it difficult to study cardiovascular and metabolic diseases; and (4) spurious links between genes and disease may be identified (GeneWatch, 2002; HGRA, 2001).

The UK Government has responded to the Committee Report by staunchly defending its position on UK Biobank and the process to develop the project. Each recommendation made by the Science and Technology Select Committee is specifically addressed (Department of Trade and Industry, 2003). In particular, the Government defends its research strategies against the charge that they are "misguided" by pointing out that "the long term strategies have been developed by the MRC council which has representation from the scientific and medical communities, in consultation with a range of organizations including the MRC research boards and government departments" (Department of Trade and Industry, 2003). The Government does acknowledge "the need to pay greater attention to communication with the research community, and to evaluation of research policy and strategy" (Department of Trade and Industry, 2003).

CARTaGENE

Researchers in Quebec are presently planning a large-scale population genetic research initiative (Cardinal & Deschenes, 2004; Network of Applied Genetic Medicine of Quebec, online). The study aims to randomly enroll 50,000 individuals (approximately 1% of the population) between the ages of 25 and 74 who will be randomly selected on the basis of their postal codes. Individuals who agree to participate will donate a blood sample and will provide demographic data, health and lifestyle information. All data collected will be anonymized. The aim is to provide a map of genetic variation in a modern heterogenous population.

The Institute for Population, Ethics and Governance (IPEG) will oversee the project. The planners of CARTaGENE are in process of securing funding for the project.

The CARTaGENE project is part of the Public Population Project in Genomics (P3G) consortium. In 2003, P3G joined three other population bank projects: the UK Biobank (United Kingdom), the GenomEUtwin (led by Finland and involving seven other countries) and the Estonian Genome Project (Estonia) International collaboration and coordination will allow CARTaGENE to work with world leaders in population genomic research (CARTaGENE, online).

Genomic Research in the African Diaspora (GRAD Study)

The first large-scale DNA and health database for research on people of African ancestry will be established at Howard University in Washington, D.C. (Kaiser, 2003).[14] The major objective of the study is "to collect, assemble and make available to the global scientific community, a well-defined and systematically characterized resource for genomic research on African Diaspora populations" (National Human Genome Center—GRAD, online).[15] Researchers hope to collect samples from 25,000 volunteers over a five-year period and to use the data to study how gene-environment interactions are implicated in common disorders affecting African-Americans. Initial recruitment will be facilitated through Howard-affiliated medical centers and doctors. Patients will later be recruited from throughout the United States and around the world. The database will be built through collaboration with Howard University and First Genetic Trust, a Chicago based company that provides secure Web-based technology for storing clinical and genetic information. A pilot project is currently being planned.

Given the history of medical experimentation on blacks in the United States, and concerns about discrimination, race-specific biobanks like the GRAD biobank will inevitably raise concerns. The relevance of race and ethnicity continue to be hotly debated subjects amongst geneticists, social scientists and legal experts. Professor Dunston, the director and founder of the National Human Genome Centre at Howard University has been described to be "in the odd position of making the scientific case against the biologic importance of race while pressing for efforts to increase inclusion of African-Americans in genetic research, both as researchers and participants" (Szalavitz, 2001).

It is hoped that race-based population research studies will be useful for studying diseases that disproportionately affect people of African ancestry; they may also help to clarify the role of genetics from socioeconomic and other environmental factors. The GRAD study is expected to yield valuable information on diseases including prostate cancer, type II diabetes, hypertension, obesity and asthma. Of the project, Francis Collins has stated "[i]t would be tragic indeed if these advances did not reach populations at particularly high risk. Thus, involving African-American populations in genetic studies on conditions such as this is critical to the future" (Szalavitz, 2001).

BIOBANKS IN THE PRIVATE SECTOR

In the United States there has been a rapid development in the number and complexity of commercial biobanks that operate to meet the growing need for human tissue, DNA and associated data in the public and private sectors (Marshall, 2001).[16] The emergence of these biobanks is seen to be a market response to the recognition of the potential and very significant value of these "products" (Andrews & Nelkin, 2001). Critics are concerned about the ethical appropriateness of the free-market approach in this field (Anderlik, 2004).

In the US there is increasing concern that protection for human research participants is in a state of disarray (Meslin, 2002).[17] Gaps in the federal regulatory system effectively make federal research guidelines inapplicable to privately funded research or commercial endeavours (Business Week, 2002; Meslin, 2002). We do not suggest that the private sector should play no role in biobanking and, in fact, the concerns over the use and sale of human biologic materials transcend the public-private distinction. For example, one news report describes the sale of neonatal blood samples by the South Carolina government to a private company for use in the development of genetic testing kits as well as to the law enforcement division for baseline studies of DNA markers (Hawkins, 2002; Trebilcock & Iacobucci, 2003).[18] This demonstrates that whether private or public entities are involved there are real opportunities for approaches that are clearly antithetical to modern ethical standards. Some concerns may, however, be specific to the private realm. For example, there are emerging concerns about the ability of commercial biobanks to adequately protect biobank participants and to return any meaningful information.

This is especially true in times of financial hardship or bankruptcy. The case of DNA Sciences and Ardais Inc. are particularly instructive.

DNA Sciences Inc.

DNA Sciences, Inc. was an applied genetics company focussed on the discovery and commercialization of DNA based diagnostic tests. Whilst operational, it sought to utilize these tests in clinical trials to understand differential responses to medications and to diagnose disease status. Through "The Gene Trust", DNA Sciences Inc. aimed to establish a database of information about individuals that included physical characteristics, health histories, and ongoing data concerning medical treatment and effectiveness. To facilitate its work, DNA Sciences utilized the internet to attract volunteers who were asked to provide contact information and a personal family health history. Once an individual was determined to be an appropriate candidate, informed consent was elicited and a blood sample was obtained and analysed. The DNA Sciences website reports that over 10,000 participants from all 50 states were registered in the Gene Trust.

As part of the recruitment strategy, the "Gene Trust Bill of Rights" (The Gene Trust Bill of Rights, online) assured participants that personally identifying genetic information would never be sold or shared with anyone outside the Gene Trust. Once collected, information would be made anonymous and the Gene Trust researchers would use only anonymous data. Genetic information would never be supplied to employers or insurance companies. Human cloning would not be undertaken, nor would DNA Sciences be associated with any such practices. Participants would be free to withdraw at any time, for any reason and without penalty.

Two provisions of the Gene Trust Bill of Rights specifically concern benefit-sharing. They provide that participants will be provided with updated information on the status of Gene Trust studies including all published material. If a participant is involved in a study that leads to the development of a new diagnostic test, that test will be provided free of charge.

On April 1, 2003, Genaissance Pharmaceuticals announced that it entered into an agreement to acquire substantially all the assets of DNA Sciences (DNA Science Press Release 2003a, 2003b). This agreement includes the Gene Trust DNA samples, anonymized medical history data as well as the computer systems that hold personally identifying data about the Gene Trust Donors (Personal Communication, Henderson, 2003). The consent form signed by Gene Trust Donors contained a provision permitting DNA Sciences to transfer samples and anonymized medical data to a third party (Personal Communication, Henderson, 2003). For business and scientific reasons it is unlikely that Genaissance will continue work on the Gene Trust. In fact, the DNA Sciences website currently states that the Gene Trust program has been discontinued and that Genaissance has no plans to use the information or biological samples that were collected (DNA Sciences, DNA Sciences "Welcome to the Gene Trust Project", online). On the website, DNA Sciences offers thanks to the gene trust donors for "[their] willingness to join the thousands of people helping [them] discover the genetic links of common diseases and conditions."

Without the de-anonymization "key" held by computer experts at DNA Sciences, Genaissance does not have the ability to decode the Gene Trust data. Questions remain as to whether the samples and associated data should be destroyed, whether the samples and data might be re-sold and whether Gene Trust donors are unavoidably and continually "at risk" for having their data used in ways to which they did not agree. Fortunately, it appears that Genaissance recognizes its continuing obligation to the Gene Trust donors. This situation, however, raises a significant number of legal and ethical questions that must be addressed.

Ardais Inc.

Ardais Inc. is a private for-profit firm located in Lexington, Massachusetts that is involved in a strategic collaboration with numerous leading United States medical institutions, including Beth Israel Deaconess Medical Center, Duke University Medical Center, Maine Medical Center and the University of Chicago. Ardais aims to develop:

> systematic large-scale procedures to comprehensively collect, process and store research-quality clinical materials and associated information; to provide these critical resources in highly optimized formats for efficient and robust design of biomedical research studies; and to support the research and clinical programs at each participating medical institution.

Ardais will license clinical materials and information for research and drug discovery purposes to qualified investigators that obtained Institutional Review Board (IRB) approval from an independent IRB.

Ardais prides itself in being a leader in the development of standards for the conduct of gene discovery and aims to continue to "identify and integrate best-demonstrated practices that will further enhance the productivity of clinical genomics research efforts worldwide." Ardais has a professional advisory committee (Best Practices Council) that is comprised of leaders from participating medical institutions and the biomedical community as well as members of the Ardais management team.

Ardais is committed to maintaining participant trust through the maintenance of high ethical standards and its commitment to ethical practices. All participation is based on free and informed consent of the participant and IRB approval is obtained at each institution. Strict data handling procedures are adhered to, including the coding of all materials and clinical information to ensure confidentiality of participant information. Through data coding, a participant's identity is inaccessible to Ardais itself and to medical researchers who license data from Ardais. The collection of clinical materials at participating medical centers does not interfere with clinical care.

Through its comprehensive website, Ardais represents to visitors that it and all participating medical institutions comply with existing federal, state and institutional requirements pertaining to the participation of human subjects in medical research, the collection and use of human biologic materials and associated information, and quality control procedures.

Public Trust and Ethical Business Practices

Corporate biobanks, including DNA Sciences and Ardais, strive to portray a corporate image based on high ethical standards and trust (Anderlik, 2004). Such an image is necessary if companies are to attract reputable scientists, sufficient investment capital, collaborators, donors and customers. It is, however, critically important to recognize that the "business" of biobanking is not a typical business. The subject matter is unique and, because of this reality, people will inevitably be outraged if firms that are involved in biobanking hold themselves out to be trustworthy but fail to meet that standard. Directors of corporate biobanks must recognize the need to abide by strict standards if the business is to be commercially successful.

Ideally, the interests of research participants should be better protected by specific regulatory mechanisms applicable to population genetic research. The case of DNA Sciences highlights the fact that bankruptcy provides unique challenges. Absent specific obligations of a trustee and utilization of the legal trust as a protective mechanism, human DNA, medical histories and personal information may be sold off as corporate assets not unlike office furniture or computer hardware (Hawkins, 2002).[19] Issues of corporate governance in relation to biobanks need to be further explored.

INTERNATIONAL HUMAN RIGHTS AND THE DUTY TO SHARE BENEFITS

Human rights have been described as "those normative principles having as their object, the benefit of the individual, which may not be infringed by the state ... or which may be achieved by political means" (Roshwald, 1958–1959). Human rights emphasize tolerance for diversity, respect for human dignity, autonomy and integrity of the person (Sieghart, 1986). The fundamental importance of "human dignity" is recognized in international human rights law and in numerous other international instruments and ethical statements. Human dignity is attributed as the source of all human rights (Knoppers, 1991).

While on the whole international instruments including treaties, conventions and covenants do not specifically address either benefit-sharing or genetic research, numerous consensus statements, codes of conduct and research guidelines do so. Importantly, there is no international treaty concerning "human genetic resources." The legal norms governing human genetic resources and benefit-sharing must, therefore, be implied by analogy and pieced together from human rights instruments, custom, general principles of law, codes of conduct, ethical standards, legal commentaries and the like.

An overview of international norms (Charter of UN, 1945; ICESCR, 1966; ICCPR, 1966; CHRBM, 1997) reveals firstly, the relevance of these human rights instruments to the understanding of the concept of benefit-sharing and secondly, the elaboration of possible meanings and applications in more "genetic-specific" instruments (CIOMS, 2002; HUGO, 1996, 2000; UNESCO, 1997). The latter documents, though not legally enforceable, reveal a recognition of the need for benefit-sharing arrangements in population genetic research.

Human Rights

The 1945 Charter of the United Nations recognizes the need for international cooperation in solving problems of an economic, social, cultural or humanitarian character. The 1948 Universal Declaration of Human Rights further stipulates the right to the basic socio-sanitary infrastructure necessary for the realization of human rights (i.e. health, security, food, clothing, etc.). It was however, the 1966 International Covenant on Economic, Social and Cultural Rights that first enunciated the concept of mutual benefit and that of freedom of research (ICESCR, 1966). It also maintained the right to benefit from scientific progress (ICESCR, 1966). That same year the International Covenant on Civil and Political Rights (1966) enshrined the concept of the need for consent to medical research.

Recently, an agency of the World Health Organization, the Council for the International Organization of the Medical Sciences, adopted ethical guidelines on biomedical research generally; they are particularly instructive (CIOMS, 2002). The Guidelines maintain that "any intervention or product developed, or knowledge generated, will be made reasonably available for the benefit of that population or community." Moreover, in addition to availability of the eventual product or knowledge, contribution to "national or local capacity to design and conduct biomedical research, and to provide scientific and ethical review and monitoring of such research are also envisioned as an example of benefit-sharing." Finally, research participants in externally sponsored research are also entitled to health care services, treatment of any ensuing injury and to benefit from services related to the interventions or products developed.

The Draft Additional Protocol to the Convention on Human Rights and Biomedicine, on Biomedical Research (Council of Europe, 2003) does not specifically address benefit-sharing. This could be due to the fact that the Convention concerns the involvement of individuals in biomedical research. The Convention itself is limited to an obligation by state parties to "take appropriate measures with a view to providing, within their jurisdiction, equitable access to health care of appropriate quality" (CHRBM, 1997).[20] Mention should also be made of the World Trade Organization's (WTO) Ministerial Declaration on the TRIPS Agreement and Public Health promoting the protection of public health and "access to medicines for all" (WTO, 2001). This statement has been translated into an agreement of WTO members to give developing countries greater access to affordable medicine by permitting them to legitimately import generic versions of expensive patented medications from countries like India and Brazil without breaching international trade laws. Supachai Panitchpakdi, director general of the WTO, commented that this "historic agreement" which allows poorer countries "to make full use of the flexibilities in the WTO's intellectual property rules in order to deal with the diseases that ravage their people ... proves once and for all that the organization can handle humanitarian as well as trade concerns" (WTO, 2003).

Whilst international human rights instruments generally invoke solidarity, international collaboration, mutual benefit and equitable access, it is these same principles that underlie the concept of benefit-sharing as revealed in the emerging norms governing genetic research.

Genetic Research

Turning then to more genetic-specific normative instruments, it was the Ethics Committee of the Human Genome Organisation (HUGO) that first introduced the concept of benefit-sharing in its 1996 Statement on the Principled Conduct of Genetic Research (HUGO, 1996). It founded its position on the credo that the human genome at the level of the species "is part of the common heritage of humanity" (HUGO, 1996). More particularly, it upheld international collaboration whilst providing "that undue inducement through compensation for individual participants, families, and populations should be prohibited" (HUGO, 1996). This prohibition, however, does not include:

> agreements with individuals, families, groups, communities or populations that foresee technology transfer, local training, joint ventures, provision of health care or of information infrastructures, reimbursement of costs, or the possible use of a percentage of any royalties for humanitarian purposes (HUGO, 2000).

In 1997, UNESCO adopted the Universal Declaration on the Human Genome and Human Rights endorsed by the United Nations. While benefit-sharing is not mentioned explicitly in this instrument, articles 18 and 19 of the Declaration stress international collaboration benefiting developing countries and the free exchange of scientific knowledge.

In its 2000 Statement on Benefit-Sharing, HUGO's Ethics Committee further clarified the meaning of "benefit-sharing" (HUGO, 2000) wherein it provided:

...all humanity share in, and have access to, the benefits of genetic research, that benefits not be limited to those individuals who participated in such research, that there be prior discussion with groups or communities on the issue of benefit sharing, that even in the absence of profits, immediate health benefits as determined by community needs could be provided, that, at a minimum, all research participants should receive information about general research outcomes and an indication of appreciation, and that profit-making entities dedicate a percentage (e.g. 1–3%) of their annual net profit to healthcare infrastructure and/or to humanitarian effort.

At the time of its publication, this HUGO Statement drew sufficient interest that an explanatory article was necessary to clarify the types of benefit-sharing models envisaged since industry and the media focussed—predictably in hindsight—on the "percentage of profits" approach (Knoppers et al., 2000). Exclusive focus on a percentage approach to the sharing of benefits severely limits the concept of benefit-sharing. Indeed, only a very small proportion of research endeavours are developed into profitable products (McClellan, 2003). With this reality in mind, the HUGO Ethics Committee envisioned the concept of "benefit" to be much broader than monetary profit. Among other things, benefits may include:

> information, therapies, improved environments (schools, libraries, sports facilities, clean water), improved health care, increased human respect ... At a minimum, participants in research should receive a thank-you (and a small token gift where the culture expects this) and information about the projects overall outcome, in understandable language ... A thank-you does not trivialize benefit-sharing. It is a sign of respect for persons and their basic intelligence and altruism (HUGO, 2000).

In October 2003 UNESCO published its International Declaration on Human Genetic Data (UNESCO, 2003)—a normative instrument that establishes the ethical principles that should govern the collection, processing, storage and use of human genetic data. With regard to the sharing of benefits, the Declaration provides that "benefits resulting from the use of human genetic data, human proteomic data or biological samples collected for medical and scientific research should be shared with the society as a whole and the international community." Benefit is defined broadly and includes, among other things special assistance to the persons and groups that have taken part in the research; access to medical care; provision of new diagnostics, facilities for new treatments or drugs stemming from the research; support for health services; capacity-building facilities for research purposes; development and strengthening of the capacity of developing countries to collect and process human genetic data, taking into consideration their specific problems; and any other form consistent with the principles set out in the Declaration.

Benefit-sharing as originally envisaged by the HUGO Ethics Committee, UNESCO and even at the level of the state as foreseen by the CBD and the Bonn Guidelines, to which we turn, should not be limited to a "partaking" of eventual (and perhaps illusory) profits.

THE PRACTICALITIES OF "BENEFIT-SHARING" IN POPULATION GENETIC RESEARCH: THE CONVENTION ON BIOLOGICAL DIVERSITY & THE BONN GUIDELINES

In this part, we turn to the key sections of the CBD (1992) and the Bonn Guidelines which, despite not being directly applicable to human genetic resources, can be used to inform the development of a benefit-sharing regime applicable for human genetic research.

Article 8(j) of the CBD provides that a member state shall:

> respect, preserve and maintain knowledge, innovations and practices of indigenous and local communities embodying traditional lifestyles relevant for the conservation and sustainable use of biological diversity and promote their wider application with the approval and involvement of the holders of such knowledge, innovations and practices and encourage the equitable sharing of the benefits arising from the utilization of such knowledge, innovations and practices.

The CBD recognizes that sovereign states hold rights over their natural resources and that inherent in this right is the authority of the state government to determine access to genetic resources. Contracting members are however obliged "to facilitate access to genetic resources for environmentally sound uses by other Contracting Parties and not to impose restrictions that run counter to the objectives of [the Convention]." Prior informed consent must be obtained from the party providing access to genetic resources, unless the approving party waives the requirement. In addition, Article 15(7) of the CBD requires that:

> Each Contracting Party shall take legislative, administrative or policy measures, as appropriate ... with the aim of sharing in a fair and equitable way the results of research and development and the benefits arising from the commercial and other utilization of genetic resources with the Contracting Party providing such resources. Such sharing shall be upon mutually agreed terms.

Importantly, the CBD also addresses the potential negative impact of intellectual property law on the CBD. Specifically, Article 16(5) provides that:

> The Contracting Parties, recognizing that patents and other intellectual property rights may have an influence on the implementation of this Convention, shall cooperate in this regard subject to national legislation and international law in order to ensure that such rights are supportive of and do not run counter to its objectives.

Pursuant to the CBD, the Bonn Guidelines are intended "to assist states in developing an overall access and benefit-sharing strategy ... and [to identify] the steps involved in the process of obtaining access to genetic resources and sharing benefits." The Guidelines make the point that stakeholder involvement "is essential to ensure the adequate development and implementation of access and benefit-sharing arrangements." "Due to the diversity of stakeholders and the reality of competing interests between stakeholders, their relative involvement must be determined on a case-by-case basis." The Guidelines suggest that member states establish a competent national authority (or authorities) that have the power to grant access to genetic resources. Such authorities may be responsible for negotiating access agreements; establishing the requirements for prior informed consent and setting mutually agreed terms; monitoring and enforcement of benefit-sharing agreements; the conservation and sustainable use of resources; mechanisms for ensuring effective stakeholder participation; and mechanisms for the effective participation of indigenous and local communities.

Pursuant to the CBD, the informed consent process should include "an indication of benefit sharing arrangements" and information about the negotiated agreement between the research partners as well as "provisions specifically addressing benefit-sharing arrangements."

In summary, the principles espoused in the international instruments and the specific mechanisms provided in the CBD and in the Bonn Guidelines are sufficiently comprehensive, flexible and logically extendable to human genetic resources. They provide a rational touchstone for the development of specific benefit-sharing guidelines applicable for human genetic resources. Inevitably and appropriately, benefit-sharing terms will be highly variable and will depend on what is regarded as fair and equitable in light of all of the circumstances of the proposed genetic research.

CONCLUSION

In light of the trend towards population genetic research and the widening gap between the developed and the developing world, mechanisms are required to ensure that the benefits of this scientific revolution can be shared equitably. The HGP provides a unique opportunity to develop a global health ethic and a "global culture of science" (Varmus, 2002).[21] The biotechnological advances derived from genomics, if applied correctly, have the potential to effect a revolutionary transformation in medicine and healthcare over the next few decades. However, if used "inappropriately and unwisely" the power of genome-related biotechnologies will inevitably exacerbate existing inequities between the developed and developing worlds (Benetar, 2003; Conference on Ethical Aspects of Research in Developing Countries, 2002). For countries with publicly funded universal healthcare systems, the integration of new genetic tests, accessible to all, will force the issue of benefit-sharing to the political front (Caulfield et al., 2003).

International human rights law, declaratory statements on the human genome, the Convention on Biological Diversity and the Bonn Guidelines provide a supportable rationale for benefit-sharing in respect of human population genetic research. We have shown that the framework outlined in the Bonn Guidelines provides detailed guidance for the implementation of workable benefit-sharing arrangements and may be used as a model to refine these arrangements in human population genetic research. Although contemplative of international research initiatives, the benefit-sharing model presented in the Bonn Guidelines can be readily integrated into the development of national or regional population genetic research initiatives. Benefit-sharing arrangements do not undermine intellectual property protection; rather, they should be modelled and viewed as a complementary tool to promote innovation, the fair sharing of the fruits of research, and public trust.

NOTES

[1] To facilitate the translation of genome-based knowledge into health benefits, the National Human Genome Research Institute has set six "grand challenges" that require it to: (1) develop robust strategies for identifying the genetic contributions to disease and drug response; (2) develop strategies to identify gene variants that contribute to good health and resistance to disease; (3) develop genome-based approaches for prediction of disease susceptibility and drug response, early detection of illness, and molecular taxonomy of disease states; (4) use new understanding of genes and pathways to develop powerful new therapeutic approaches to disease; (5) investigate how genetic risk information is conveyed in clinical settings, how that information influences health strategies and behaviors, and how these affect health outcomes and costs; and (6) develop genome based tools that improve the health of all.

[2] The difficulties include: (1) defining biologically valid phenotypes; (2) identifying and quantifying environmental exposures; (3) generating sufficient and useful genotypic information; and (4) studying human subjects.

[3] Large-scale projects include, among others, the Icelandic Health Sector Database, the Estonian Genome Project, Biobank UK, Marshfield Clinic's Personalized Medicine Research Project (U.S.), the Latvian Genome Project, Cartagene (Quebec), ProgeNIA (Sardinia), UmanGenomics (Sweden) and the Genomic Research in the African Diaspora (US) (see Austin et al., 2003).

[4] For a word of caution, see Haga et al. (2003).

[5] See Final Act Embodying the Results of the Uruguay Round of the Multilateral Negotiations, Marrakesh Agreement Establishing the World Trade Organization, signed at Marrakesh (Morocco), April 15, 1994, Annex 1C, Agreement on Trade-Related Aspects of Intellectual Property Rights, GATT, Doc.MTN/FA/Add.1 (December 15, 1993); reprinted in The Results of the Uruguay Round of Multi-lateral Trade Negotiations - The Legal Text, vol. 31 (GATT Secretariat, ed., 1994) 365–403; 33 I.L.M. 1197, 1200; 25 I.I.C. 209. The TRIPS Agreement came into force on January 1, 1995. It sets out minimum national standards for the protection of intellectual property with which members of the World Trade Organization must comply. The TRIPS Agreement is generally reflective of the strong IP regimes favored by the North. Developed countries were given one year to comply whilst developing nations were given until January 1, 2000. For countries required to extend patent protection to new areas such as pharmaceuticals, a further five years was granted before protection had to be introduced. Least Developed Countries were expected to enact TRIPs by 2006 but the Doha Ministerial Declaration on the TRIPs Agreement and Public Health allows them until 2016 to implement protection of pharmaceutical products.

[6] The defendant's motion to dismiss the case was denied on July 8, 2002. The case has been transferred to the United States District Court for the Southern District of Florida, Miami Division, number 02-022244-CIV-MORENO. On May 29, 2003, Judge Moreno ordered that the case will proceed to trial (*see Greenberg* et al. v. *Miami Children's Hospital and Reuben Matalon*)

⁷ An unfortunate situation arose in Newfoundland when researchers from Baylor College of Medicine in Texas flew into St. John's to study a family with a high incidence of ARVC, a congenital heart disease that renders victims susceptible to cardiac arrest at an early age. Researchers spent the weekend in Newfoundland collecting blood samples from family members. Follow-up treatment and genetic counselling were never offered. Local physicians responsible for caring for the family were not given access to the data and participants were given no indication as to whether or not they were at risk.

⁸ The US Office for Human Research Protections [OHRP] completed an investigation in response to a complaint that a Harvard School of Public Health faculty member violated processes and regulations with respect to genetic research carried out in rural Anhui province in China. In a determination letter dated March 28, 2002, the OHRP concluded that there were procedural lapses but did not find any harm to any participants in the studies and did not apply sanctions to the university. The two researchers involved were reprimanded for omissions in fulfilling obligations pursuant to policies and rules and for failing to seek timely review and approval of studies. The problems have since been corrected and research continues.

⁹ In the United States, for example, the relative contribution of industrial involvement in biomedical research grew from approximately 32% in 1980 to 62% in 2000 (Bekelman et al, 2003, Statistics Canada, 2002a). Of the total $4.2 billion spent in 2001 on health research performed by universities, teaching hospitals, business enterprises, government labs, and private non-profit organizations, only 22.4% was provided by government (federal or provincial) (Statistics Canada, 2002b). In 1999, three of ten biotechnology firms in Canada were spin-off companies. Ninety-one percent (91%) were formed by universities or teaching hospitals; 75% focus on human health.

¹⁰ Article 2 of the Act states that the legislation extends to the creation and operation of a centralized health sector database and does not apply to the storage or handling of or access to, biologic samples.

¹¹ The Icelandic Medical Association and the Mannevernd (the Association of Icelanders for Ethical Science) both opposed the bill. Criticisms have been made over issues of privacy, confidentiality and the inevitable erosion of doctor-patient trust. Additionally, there is a belief that researchers, other than those associated with deCODE or Hoffman-La Roche will have unequal access to the data (see Andersen and Arnason, 1999).

¹² It is reported in this article that genotype data from Estonians and other Caucasians demonstrate that there are only minor differences between European populations. The implication is that research done in Estonia will be widely applicable to the whole of the European population.

¹³ GeneWatch is a public interest group based in the United Kingdom that aims to see genetic technologies developed responsibly and used to promote human health and to safeguard the environment. GeneWatch also hopes to engage the public in the decisions concerning genetic technologies and to support research on the impact of genetic technologies. GeneWatch asserts that Biobank UK (and other biobanks) raise issues that, to date, have not been adequately addressed in research policy. They feel it is unlikely that guidelines developed to regulate professionals will be sufficient to control the realities of market-driven relationships that will inevitably arise. GeneWatch recommends the enactment of new legislation, regulations or guidelines to ensure that these relationships are properly controlled.

¹⁴ Details of the project were announced in June 2003. Howard University is described on its website as: "a comprehensive, research oriented, historically Black private university providing an educational experience of exceptional quality to students of high academic potential with particular emphasis upon the provision of educational opportunities to promising Black students." Online: Howard University <www.founders.howard.edu/presidentReports/Mission.htm>.

¹⁵ Specifically, the study will: (1) identify genetic polymorphisms in African Americans and their ancestral populations; (2) perform molecular phenotyping of genotypically defined biological specimens; (3) establish complementary epidemiological databases; and (4) develop statistical genetic models for analyses of families and gene flow among populations.

¹⁶ This article focuses on the business strategy of the First Genetic Trust Inc. which describes itself as "an intermediary between patients and researchers." Individuals permit the company to store their genetic information in a confidential database for use in clinical research; visit their corporate website at: <www.firstgenetic.net/>.

[17] In the United States, numerous federal agencies as well as public and private institutions that do not receive federal funding are not subject to the Common Rule. A similar loophole exists in Canada wherein entities that do not receive federal funding are not required to comply with the Tri-Council Policy Statement.

[18] The authors note that too often people conclude that flaws in the private market imply the need to maintain public sector influence. They propose that in discussing the merits of public and private action, the analysis must be relative. They also suggest that private market shortcomings "pale in comparison to the flaws associated with public provision or public oversight of private actors."

[19] This article notes that when the Belgian biobank "Spitters.com" went bankrupt, 500 saliva swabs donated for research were put up for sale alongside the office equipment.

[20] The 43 member States must sign and ratify this Additional Protocol. Only those countries that have previously signed and ratified the Convention can adopt the Additional Protocol. Countries such as Canada can be invited to sign and ratify the Convention and its Protocols.

[21] Varmus advocates the creation of "a Global Science Corps" wherein scientists from developed countries would go into the world's poorest countries and use local talents to work towards local scientific goals. Varmus asserts that September 11, 2001 alerted us to the resentments that have arisen as a result of economic asymmetries and that something must be done about it to ensure that "over the longer haul, we can put the science we value so highly to work to help sustain improved health."

REFERENCES

Anderlik MR (2004) Commercial biobanks and genetic research: banking without checks? In: Knoppers BM (ed) Populations and genetics: legal and socio-ethical perspectives. Martinus Nijhoff, Boston

Andersen B, Arnason E (1999, June 5) Icelands's database is ethically questionable. BMJ 318:1565

Andrews L, Nelkin D (2001) Body bazaar: the market for human tissue in the biotechnology age. Crown Publishers, New York

Annas G J (1993) Privacy rules for DNA databanks: protecting coded "future diaries. JAMA 270(19):2346–2350

Anonymous (1995, July 15) Western eyes on China's eugenics law. Lancet 346(8968):131

Austin MA, Harding S, McElroy C (2003) Genebanks: a comparison of eight proposed international genetic databases. Community Genetics, Community Genet 6(1):37–45

Barton JH (1997) Patents and antitrust: a rethinking in light of patent breadth and sequential innovation Antitrust Law Journal, Anti LJ 65(2):449–466

Baslar K (1998) The concept of the common heritage of mankind. : Martinum Nijhoff Publishers, Boston

Bekelman JE, Lie Y, Gross CP (2003) Scope and impact of financial conflicts of interest in biomedical research: a systematic review JAMA 289:454

Benatar SR, Daar AS, Singer PA (2003) Global health ethics: the rationale for mutual caring. International Affairs 79(1):107–138, at 109 citing Singer PA, Daar AS (2001, October 5) Harnessing genomics and biotechnology to improve global health equity. Science 294: 87–89

Berg K (2001) The ethics of benefit-sharing. Clin Genet 59(4): 240–243

Berger A (1999, January 2) Private company wins right to icelandic database. BMJ 318(7175):11

Beskow LM et al (2001) Informed consent for population-based research involving genetics. JAMA 286(18):2315–2321

Biobank UK. Draft Protocol, available online: BioBank UK <http://www.ukbiobank.ac.uk/documents/draft_protocol.pdf>

Blumenthal D (1996, December) Ethics issues in academic-industry relationships. Academic Medicine 71:1291–1296

Business Week (2002, April1 5) Balancing privacy and Biotechnology. Editorial *Business Week*

Cardinal G, Deschenes M (2004) Surveying the population biobankers. In: Knoppers BM (ed) Populations and genetics: legal and socio-ethical perspectives. Martinus Nijhoff, Boston, pp 37–94

CARTaGENE: http://www.cartagene.qc.ca/

Cassese A (1990) International law in a divided world. Clarendon, Oxfordat 391

Caulfield TA et al (2003) Genetic technologies, health care policy and the patent bargain. Clin Genet 63(1):15–18
Caulfield T (2002) Gene banks and blanket consent. Nat Rev Genet 3(8):577
Chakravarti A (1999) Population genetics – making sense out of sequence. Nat Genet 21(Suppl.) 56
Charter of the United Nations (1945, June 26) Can. T.S. 1945 No. 7.
Cho MK et al (2003) Effects of patents and licenses on the provision of clinical genetic testing services. J Mol Diagn 5(1):3–8
Collins FS, Green ED, Guttmacher AE (2003) A vision for the future of genomics research. Nature 422(6934): 835–847
Commission on Intellectual Property Rights (2002) Integrating intellectual property rights and development policy. CIPR, London) at 11 citing Mansfield E (1986) Patents and Innovation. Manage Sci 32(2):173–181
Convention on Biological Diversity (CBD) (1992, June 5) 31 I.L.M. 818
Council for International Organizations of Medical Sciences (CIOMS) (2002) International ethical guidelines for Biomedical research involving human subjects. CIOMS, Geneva.online: CIOMS < www.cioms.ch/frame_guidelines_nov_2002>
Council of Europe (1997, April 4) Convention on Human Rights and Biomedicine (CHRBM). ETS No. 164
Council of Europe (2003) Steering committee on bioethics, draft additional protocol to the convention on human rights and Biomedicine, on Biomedical Research, CDBI/INF 6
Dawson E et al (2002) A first-generation linkage disequilibrium map of human chromosome 22. Nature 418(6897):544–548.
deCODE Genetics, An Informed Consent for Participation in a Genetic Study of (name of Disease) online: deCODE Genetics <www.decode.com>
Department of Trade and Industry (2003, June) Government Response to 'The work of the Medical Research Council' Report by the House of Commons Science and Technology Select Committee (HC 132). online: UK Parliament <www.ost.gov.uk/research/councils/govt_response_to_select_committee_report_on_mrc.pdf>
Deschenes M et al (2001) Human genetics research DNA banking and consent: a question of 'form'? Clin Genet 59(4):221–239
DNA Sciences "Welcome to the Gene Trust Project" online: DNA Sciences <http://www.dna.com/sectionHome/sectionHome_TheDNASciencesGeneTrustProject.asp>
DNA Sciences, Press release (2003a, April 1) Genaissance pharmaceuticals enters into agreement to acquire assets of DNA sciences. online: DNA Sciences www.dna.com/pressRelease/pressRelease.jsp?site=dna&link=20030331.htm
DNA Sciences, Press Release (2003b, May 12) Genaissance pharmaceuticals' acquisition of substantially all of the assets of DNA sciences is approved. online: DNA Science <www.dna.com/pressRelease/pressRelease.jsp?site=dna&link=20030512.htm>
Eisenberg RS (1992) Genes, patents and product development. Science 257(5072):903–908
Estonian Genome Project, available online: Eesti Geenivaramu <http://www.geenivaramu.ee/mp3/trykisENG.pdf> at 5
Estonia, Human Genes Research Act (HGRA) RT I 2000, 104, 65, s.12 (entered into force 8 January 2001) online: Estonian Genome Project
Gana RL (1995) Has creativity died in the third world? some implications of the internationalization of intellectual property. Denver J Int Law Policy 24(1):109–144
Gandi I Address (Lecture presented to the World Health Assembly, 1982), cited in Gadbaw RM,Kenny LA, *"India"* in Intellectual Property Rights: Global Consensus, Global Conflict? Gadbaw RM, Richards TJ (eds) (1988) Westview Press, Boulder,)
GeneWatch UK (2002, November) Biobank UK – A Good Research Priority? Parliamentary Briefing No. 3. available online: GeneWatch <http://www.genewatch.org/HumanGen/Publications/MP_Briefs.htm>
Gold ER (1996) Body parts: Property rights and the ownership of human biological materials. Georgetown University Press, Washington)

Gostin LO, Hodge JG (1999) Genetic privacy and the law: an end to genetic exceptionalism Jurimetrics 40(1):21–58
Greely HT (1997) The control of genetic research: involving the 'groups between' Hous L Rev 33:1397
Greely HT (2000) Iceland's plan for genomics research: facts and implications. Jurimetrics 40(2): 153–191
Habeck M (2002, October 17) Estonia jumps on the gene bank train. The Scientist (). online: The Scientist.com <www.biomedcentral.com/news/20021017/08>
Haga SB, Khoury MJ, Burke W (2003) Genomic profiling to promote a healthy lifestyle: not ready for prime time. Nat Genet 34(4):347–350
Hamilton MA (1997) The TRIPs agreement: imperialistic, outdated, and overprotective. In: Moore AD (ed) Intellectual property: moral, legal and international dilemmas. Rowman & Littlefield, Boulder
Harrison CH (2002) Neither moore nor the market: alternative models for compensating contributors of human tissue. Am J Law Med 28(1):77–105
Harvard School of Public Health (2003, May 30) Conclusion of U.S. Government's Inquiry into HSPH Genetic Research in China. available online: Harvard University <www.hsph.harvard.edu/press/releases/pres05302003.html>
Hawkins D (2002, February 12)Keeping secrets. U.S. News & World Report
House of Commons Science and Technology Committee, "The Work of the Medical Research Council: Third Report of Session 2002–03" available online: UK Parliament <www.publications.parliament.uk/pa/cm200203/cmselect/cmsctech/132/132.pdf> at 28
Human Genetics Commission (2002) Inside information: balancing interests in the use of personal genetic data. A Summary Report by the Human Genetics Commission (2002, May). Available online: http://www.hgc.gov.uk/ UploadDocs/DocPub/Document/insideinformation_summary.pdf
Human Genome Organisation Ethics Committee (2000) HUGO Statement on Benefit-Sharing (SBS) Genome Digest 6:7–9
Human Genome Organisation Ethics Committee (1996) HUGO Statement on the Principled Conduct of Genetic Research (PCGR) Genome Digest 3:2–3
Iceland Act on a Health Sector Database, no.139/1998 (Passed by Parliament 123rd Sess., 1998–99). Article 2 Iceland, Ministry of Health and Social Security <http://www.raduneyti.is/interpro/htr/htr.nsf/pages/lawsandregs>
Iceland Act on Biobanks, no.110/2000. Both Acts are available online: Iceland, Ministry of Health and Social Security <http://www.raduneyti.is/interpro/htr/htr.nsf/pages/lawsandregs>
Juengst ET (1998) Group identity and human diversity: keeping biology straight from culture. Am J Hum Genet 63(3):673–677
Kaiser J (2003) African-American population biobank proposed. Science 300(5625):1485
Kaiser J (2002) Population databases boom, from Iceland to the U.S. Science 298(5596):1158–1161
Knoppers BM et al (2000) "HUGO Urges Genetic Benefit-Sharing". Community Genet 3(2):88–92
Knoppers BM (2000[1997]) Population genetics and benefit sharing. Community Genet. 3(4):212–214
Knoppers BM (1999) Biotechnology: sovereignty and sharing. In: Caulfield TA, Williams-Jones B (eds) The commercialization of genetic research: ethical, legal and policy issues. Kluwer Academic/Plenum Publishers, New York, pp 1–11
Knoppers BM (1991) Human dignity and genetic heritage. Law Reform Commission of Canada, Ottawa
Litman M, Robertson G (1996) The common law status of genetic material. In: Knoppers BM (ed) Legal rights and human genetic material. Emond Montgomery Publications Ltd., Toronto, pp 51–81
Lowrance WW (2001) The promise of human genetic databases: high ethical as well as scientific standards are needed BMJ 322(7293):1009–1010
Markel H (1992) The stigma of disease: implications of genetic screening Am J Med 93:209
Marshall E (2001) Company plans to bank human DNA profiles Science 291(5504):575
Mathew C (2001) Postgenomic technologies: hunting the genes for common disorders. BMJ 322(7293):1031–1034
McClellan MB (2003, June 23) Lecture, BIO 2003, Washington DC. available online: BIO 2003 <http://www.bio.org/events/2003/media/mcclellan_0623.asp>
McInnis MG (1999) The assent of a nation: genethics and Iceland. Clinical Genetics 55(4):234–239

Merges RP, Nelson RR (1990) On the complex economics of patent scope Colum L Rev 90:839
Meslin EM (2002) Raising the bar in research ethics: traditional obligations are not enough. Postgraduate Medicine 112:5
National Human Genome Center, Genomic Research in the African Diaspora (GRAD) <www.genomecentre.howard.edu/grad.html>
Network of Applied Genetic Medicine of Quebec (FRSQ), "Cartagene in Brief" available online: RMGA www.rmga.qc.ca
Participants in the 2001 Conference on Ethical Aspects of Research in Developing Countries (2002) Fair benefits for research in developing countries. Science 298(5601):2133–2134
Pennisi E (1998) Brazil wants cut of its biological bounty. Science 279(5356):1445
Pomfret J, Nelson D (2000, December 20) In Rural China, a genetic mother lode. Washington Post, A1
Robertson JA (1999) Privacy issues in second stage genomics. Jurimetrics 40(1):59–76
Roshwald M (1958–59) The concept of human rights. Philosophy and Phenomenological Research 19(3):354–379
Shiva V (1999, November) The real costs of globalization. (Lecture presented to The Parkland Institute's 3rd Annual Conference, "The Corporation as Big Brother: Challenging the Privatization of Knowledge".) Edmonton, Alberta
Shiva V (1997)Biopiracy: the plunder of nature and knowledge. Between the Lines, Toronto
Sieghart P (1986) The lawful rights of mankind. Oxford University Press, New York:) at 42
Staples S (2000) Human resource: Newfoundland's 300-year-old genetic legacy has triggered gold rush. Report on Business 17(3):117–120
Statistics Canada (2002b) A profile of spin-off firms in the biotechnology sector. Innovation Analysis Bulletin 4(1):12
Statistics Canada (2002a) Estimates of total expenditures on research and development in the health fields in Canada, 1988 to 2001. available online: Statistics Canada <www.statcan.ca/cgi-bin/downpub/listpub.cgi?catno=88F0006XIE2002007>
Szalavitz M (2001) Race and the genome. online: Howard University Human Genome Center <www.genomecenter.howard.edu/article.htm>
The Bonn Guidelines on Access to Genetic Resources and Fair and Equitable Sharing of the Benefits Arising out of their Utilization (2002) UN Doc UNEP/CBD/COP/6/20 online: Convention on Biological Diversity <www.biodiv.org/decisions/default.asp?m=cop-06&d=24>
The Gene Trust Bill of Rights online: DNA Sciences, <http://www.dna.com/sectionHome/SectionHome.jsp?site=dna&link=The GeneTrustBillofRights.htm>
The International Covenant on Economic, Social and Cultural Rights (ICESCR) (1966) G.A. Res. 2200, U.N. GAOR, 21st Sess., Supp. No. 16 at 49, U.N. Doc. A/6316; International Covenant on Civil and Political Rights (ICCPR), 19 December 1966, 999 U.N. T.S. 171, arts. 9–14, Can. T.S. 1976 No. 47, 6 I.L.M. 368
Trebilcock MJ, Iacobucci EM (2003) Privatization and accountability Harv Law Rev 116(5):1422–1453
United Nations Educational, Scientific and Cultural Organisation (UNESCO) (1997) Universal Declaration on the Human Genome and Human Rights (UDHG), 29th Sess., 29 C/Resolution 19
United Nations Educational, Scientific and Cultural Organisation (UNESCO) (2003), International Declaration on Human Genetic Data, 32nd Sess., 32 C/29 Add.2. The text of the declaration as finalized by the working group of Commission III on 7 October 2003 is included as Section III of this document
United Nations Convention on the Law of the Sea (1982) UN Doc. A/CONF.62/122 art. 136
Universal Declaration of Human Rights (UDHR) (1948) GA Res. 217(III), UN GAOR, 3d Sess., Supp. No. 13, UN Doc. A/810
Varmus (2002) Harold building a global culture of science. Lancet 360(Suppl):1
Weijer C (2000) Benefit sharing and other protections for communities in genetic research. Clinical Geneti 58(5):367–368
Weijer C, Goldsand G, Emanuel EJ (1999) Protecting communities in research: current guidelines and limits of extrapolation. Nat Genet 23(3):275–280
Wertz DC, Fletcher JC (1998) Ethical and social issues in prenatal sex selection: a survey of geneticists in 37 nations. Soc Sci Med 46(2):255–273

Wertz DC (2002) Did eugenics ever die? Nat Rev Genet 3(6):408
World Health Organization (2002) Advisory Committee on Health Research, Genomics and World Health Organization World Health, Genomics and World Health. World Health Organization, Geneva)
World Trade Organization (Press Release), "Decision Removes Final Patent Obstacle to Cheap Drug Imports" 30 August 2003, online WTO: <www.wto.org/english/news_e/pres03_e/pr350_e.htm>
World Trade Organization Declaration on the TRIPS Agreement and Public Health (Doha Statement), 2001, Doc. WT/MIN(01)/Dec/2, available online: http://www.wto.org/english/thewto_e/minist_e/min01_e/mindecl_trips_e.htm

CASES

Moore v. *Regents of the University of California* 249 Cal. 494 (Ct. App. 1988)
271 Cal. 146 at 150, 793 P. 2d 479 (1990) cert. Denied, 111 S. Ct. 1388 (1991). Compliant (N.D. Ill. Oct. 30, 2000) (No. 000CV-6779).
Institute for Science, Law and Technology, "Background Information on *Greenberg* v. *Miami Children's Hospital* et al., online: Chicago-Kent College of Law <http://www.kentlaw.edu/islt/canavanupdate.html
Daniel Greenberg et al v. *Miami Children's Hospital and Reuben Matalon*, 26F.Supp.2d 1064, 16 Fla. L. Weekly Fed. D417.

JORGÉ CABRERA MEDAGLIA[1]

CHAPTER 9

BIOPROCESSING PARTNERSHIPS IN PRACTICE: A DECADE OF EXPERIENCES AT INBio IN COSTA RICA

INTRODUCTION

The importance of biotechnology for food, agriculture, human health, environmental protection, etc., has been outlined by diverse studies and emphasized by entities such as the Food and Agriculture Organization of the United Nations and the United Nations Environment Programme. At the same time, accessing and acquiring these technologies is especially complex due to their proprietary character (patents, plant breeders' rights, and other intellectual property rights). In the great majority of cases, big transnational firms own these rights. This is because they have the financial capacity to allocate R&D resources for new products and biotechnological processes.[2]

In order to close the gap between those who control these technologies and those who need them, many different schemes have tried to facilitate the access and transfer of biotechnology, mostly in agriculture. One of the most well known is the programme of the International Service for the Acquisition of Agri-biotechnologies (ISAAA) (Krattiger, 2000). Another interesting attempt has been tried in Costa Rica by the National Biodiversity Institute (INBio). Through agreements on access and the supply of biodiversity (samples and extracts), important technology has been acquired (not always involving biotechnology) that has helped to consolidate a basic infrastructure that can add value to resources and facilitates the discovery of new intelligent uses for genetic resources. As both a private institution of public interest and a non-profit charter, INBio can share important experiences about how to spread benefits derived from access to genetic resources.

INBio's experience illustrates how the objectives of the Convention on Biological Diversity can be achieved by sharing the benefits derived from access to genetic resources, including the transfer of technology. It also reveals the importance of the collaborative agreements that allow our countries to access the technology and know-how needed to add value to biodiversity. Such agreements contribute to the conservation and sustainable use of our biodiversity and thereby improve our quality of life.

INBIO'S EXPERIENCE

This section presents a brief summary of the most important, successful and promising agreements by INBio. The information is based on data provided by INBio's Bioprospecting Unit. INBio has engaged in an array of relationships over the past 15 years with a variety of large multinational corporations and small entrepreneurial firms.

Research Collaboration Agreements with Industry

INBio-Merck: Search for sustainable uses of Costa Rican biodiversity This was the first agreement signed with a commercial company (October of 1991). It allowed Merck to search for sustainable uses of Costa Rican biodiversity of interest to the pharmaceutical industry and veterinary science. It was renewed in 1994, 1996, and 1998 on similar terms. Studies to determine the potential use of a limited number of extracts of plants, insects, and environmental samples have been completed, and the agreement has given INBio access to technology, teams, and training. This collaboration has ended and no product reached the market.

Chemical prospecting in a Costa Rican conservation area This project began in 1993 and ended in September 1999. Financed by the United States' National Institutes of Health (NIH), it was one of the world's five International Groups of Cooperation in Biodiversity (ICBG). Located in the Guanacaste Conservation Area, collaborators included the University of Costa Rica, Cornell University, and Bristol Myers Squibb. It examined tropical insects for possible pharmaceutical products and increased the local human resource capacity in ecology, taxonomy, and ecochemistry.

INBio-Givaudan Roure: Fragrances and aromas Constantly searching for new ways to utilize our biodiversity, in 1995 INBio began to explore its potential for fragrances and aromas with the company Givaudan-Roure. Aromas and fragrances were taken directly from forest air surrounding fragrant objects. The objective was to determine whether new products could be generated and to investigate technology transfer options in this area. A royalty rate was established, and the agreement concluded its activities in Costa Rica in the middle of 1998.

INBio-BTG-Ecos La Pacífica In agriculture, INBio seeks to integrate bioprospecting discoveries with the country's economic development. This process began with the 1992 INBio-British Technology Group (BTG) Agreement, which allowed INBio to begin the investigation, characterization, and manufacture of a product with nematicidal activity derived from a tree found in Costa Rica's dry tropical forest. At the same time, investigations were developed jointly with the corporation of Ecos La Pacífica to determine the species' growing conditions, the effectiveness of DMDP in tropical crops, and its production methods. Greenhouse and field trials began in 1999; they continue to-date with very satisfactory results. BTG has paid a small amount of money to both INBio and Ecos for licensing a patent related to the use of DMDP.

INBio-Diversa: Search for enzymes from extremophilic organisms with chemical industry application To explore new enzymes discovered in aquatic and terrestrial microorganisms that live under extreme conditions, INBio signed a research agreement with the DIVERSA biotechnical firm in 1995. Renewed in 1998 and 2002, the agreement gathers bacteria from different Conservation Areas of Costa Rica in order to identify and isolate novel, useful enzymes for industry. The agreement also guarantees training for Costa Rican scientists in collection methods, isolation, and molecular biology, specifically in cloning and characterizing genes associated with the enzymes.

INBio-INDENA S.P.A.: Search for compounds with antimicrobial and antiviral activity To obtain compounds with antimicrobial potential for use as active ingredients in cosmetics, INBio and the phytopharmaceutical company INDENA (Milan, Italy) signed a collaboration agreement in 1996. Extracts from selected plants are evaluated in bioassays to determine their antimicrobial activity. The final process is carried out at INDENA. A second phase of the project began in 2000.

INBio-Phytera Inc. Traditionally drugs have been developed from the extracts of leaves, roots, bark, and other plant parts. Advances in biotechnology now make it possible to derive them by cultivating cells, which only requires extremely small samples and can produce a diversity of chemical substances—more than when the original plant is used. In 1998 INBio signed an Agreement with Phytera Inc. to pursue this process. It concluded in 2000.

INBio- Eli Lilly: Search for new compounds This project ran from 1999 to 2000 and searched for botanical compounds with pharmaceutical applications. As a result of this successful collaboration, Ely Lily donated to INBio technology to prepare fractions (Bioexplore), which allowed INBio to provide fraction services and improve its research and development capabilities.

INBio-Akkadix corporation: Search for compounds with nematicidal activity This project was carried out with the Akkadix Corporation from 1999 to 2001. Its main objective was to search for alternatives to controlling nematodes.

Agreements with Academia

INBio also has academic investigation agreements with national and international universities. These vary in focus but are all geared toward producing knowledge about our biodiversity, discovering potential novel solutions to current problems in a number of fields, and developing new products.

INBio-University of Strathclyde This agreement provided access to new technologies and methodologies and also established connections with the Japanese private sector. INBio supplied a limited number of plant extracts for evaluation by several Japanese industries. This agreement existed from 1997 to 2000.

INBio-University of Massachusetts: Search for potentialiinsecticides INBio's collaboration with the University of Massachusetts to look for compounds with insecticidal activity was made possible through the support of the NIH. This investigation began in October 1995 and concluded in 1998. It performed enzymatic bioassays on extracts of plants, insects, bryophytes, and mollusks.

INBio-University of Guelph: Development of new technologies for medicines based on plants, an international interdisciplinary initiative This agreement with the University of Guelph was signed in 2000 and extended to 2003. The main objective was to search for new pharmaceutical products through such techniques as tissue cultivation.

Other Types of Agreements

Validation of promissory plants This project was financed by the CR-USA Foundation. It contemplated three sub-projects to obtain information that could improve the quality of life in Costa Rica. In collaboration with the Center for Research and Diagnosis in Parasitologia of the University of Costa Rica (CIDPA), two plants were studied to isolate their active components against malaria. This investigation gave continuity to the excellent results of the ICBG project.

Also, in collaboration with the UME (Unit of Electronic Microscopy), LEBI (Laboratory of Biological Assays), and the National Children's Hospital, plants traditionally used to treat gastritis were validated by their anti-helicobacter pylori activity. To explore their economic feasibility, some species were also validated by alkaloid content.

The Chagas project INBio joined with EARTH, the National University of Costa Rica and other Latin American institutions in Brazil, Mexico, Chile, Argentina, Uruguay, and the United States (NASA) as part of "The ChagaSpace Project." This project sought a solution to one of the most serious public health problems in Latin America: Chagas disease (American Tripanosomiasis). In 1997 INBio researched plants with inhibitory activity for this disease. In 2001 the United States Congress approved a fund dedicated to financing this project again, and work on the bioassays has been restarted.

INBio-IDB: Program from support of the development of the use of the biodiversity by small enterprises In February of 1999, INBio signed an agreement with the Interamerican Development Bank to formalize the terms of a grant for non-reimbursable technical cooperation to supports biodiversity development for small companies. Six projects have been approved: (1) Agrobiot S.A.: Propagation of Costa Rican tropical plants for commercialization as eco-educational souvenirs; (2) Laboratorios Lisan S.A.: Pharmaceutical products based on medicinal plants. At least five natural products will be commercialized in 2004 and 2005; (3) La Gavilana: Development of a model for eco-friendly practices in vanilla production through the identification of a biopesticide that allows for the organic production

TABLE 1. Most significant Research Collaborative Agreements with Industry and Academia (from 1991 to 2002)

Industrial or academic partner	Natural resources accessed/objectives	Field of primary application	Research activities in Costa Rica
Cornell University	INBio's capacity building	Chemical Prospecting	1990–1992
Merck & Co.	Plants, insects, micro organisms	Human health and veterinary	1991–1999
British Technology Group	DMDP, compound with nematocidal activity*	Agriculture	1992–present
ECOS	*Lonchocarpus felipei*, source of DMDP*	Agriculture	1993–present
Cornell University and NIH	Insects	Human health	1993–1999
Bristol Myers & Squibb	Insects	Human health	1994–1998
Givaudan Roure	Plants	Fragrances and essences	1995–1998
University of Massachusetts	Plants and insects	Insecticidal components	1995–1998
Diversa	DNA from Bacteria	Enzymes of industrial applications	1995–present
INDENA SPA	Plants*	Human health	1996–present
Phytera Inc.	Plants	Human health	1998–2000
Strathclyde University	Plants	Human health	1997–2000
Eli Lilly	Plants	Human health and agriculture	1999–2000
Akkadix Corporation	Bacteria	Nematocidal proteins	1999–2001
Follajes Ticos	Plants	Ornamental applications	2000–present
La Gavilana S.A.	*Trichoderma* spp*	Ecological control of pathogens of *Vanilla*	2000–present
Laboratorios Lisan S.A.	None*	Production of standardized phytopharmaceuticals	2000–present
Bouganvillea S.A.	None*	Production of standardized biopesticide	2000–present
Agrobiot S.A.	Plants*	Ornamental applications	2000–present
Guelph University	Plants*	Agriculture and Conservation purposes	2000–present
Florida Ice & Farm	None*	Technical and Scientific support	2001–present
Chagas Space Program	Plants, fungi*	Chagas disease	2001–present
SACRO	Plants*	Ornamental applications	2002–

* These agreements include a significant component of technical and scientific support from INBio.
Source: Tamayo et al. forthcoming in 2004.

of vanilla; (4) Industrias Caraito S.A.: Generated added value for the Carao agro-industry; (5) Bougainvillea S.A.: Research to develop and produce a Biocide from Quassia amara wood and (6) Follajes Ticos S.A.: Ornamental plants native to the forest and with potential for successful commercialization: Several new species are under domestication.

The significance of this contract approach must not be underestimated. These contractual arrangements have made possible different joint initiatives (e.g., the Cooperative Biodiversity Groups, etc.) as well as studies on the effects of benefit sharing. Table 1 summarizes the main collaborative agreements.

MAIN BENEFITS AND RESEARCH RESULTS

These and other contract relationships have provided great benefits to INBio and Costa Rica, including: monetary benefits through direct payments; payment for supplied samples; funds for research budgets; the transfer of important technology enabling infrastructure development at the Institute (biotechnology lab, etc.) that can be used to investigate and generate its own products; training of scientists and experts in state-of-the-art technology; negotiation experience, particularly knowledge of the market and of the probabilities of identifying biodiversity resources for intellectual use; support of conservation through payments made to the Ministry of the Environment to strengthen the National System of Conservation Areas; transfer of equipment to other institutions, such as the University of Costa Rica; future royalties and milestone payments to be shared 50:50 with the Ministry of the Environment; and establishment of national capabilities for assessing the value of biodiversity resources.

Many of these projects promise to offer longer term benefits, as their research results (Table 2) are tested for commercial application over the coming years. Overall, these projects contributed significantly to biodiversity conservation in the 1993–2000 period (Table 3).

Legal considerations

In Costa Rica, genetic and biochemical resources are the property of the State, with qualifications regarding public goods. In the case of indigenous territories and the public or private ownership of the lands or biological resources containing the genetic and biochemical resources, the owners' prior informed consent is required for access. But this does not grant them a right of property for the genetic and biochemical components. The law requires the applicant to attach the prior informed consent for access granted by the owner of the land, by the authority of the indigenous community, or by the Director of the Area of Conservation (Article 65, Law of Biodiversity).

Costa Rica's Law of Biodiversity No. 7788 of 27 May 1998 applies to biodiversity components that are under the State's sovereignty and to the processes and activities carried out under its jurisdiction or control, independently from those effects manifested inside or outside of the national jurisdiction. This Law specifically regulates the use, management, associated knowledge, and sharing of the benefits and derived costs of utilizing biodiversity components (Article 3). Also,

TABLE 2. Outputs Generated Since 1992 as a Result of Research Collaboration Agreements with INBio

Project	Year Initiated	Major output
Merck & Co.	1992	27 patents
BTG/ECOS	1992	DMDP on its way to commercialisation
NCI	1999	Secondary screening for anti- cancer compounds
Givaudan Roure	1995	None yet
INDENA	1996	Two compounds with significant anti-bacterial activity
Diversa	1998	Two potential products at initial stages/Publication underway
Phytera Inc.	1998	None yet
Eli Lilly & Co.	1999	None yet
Akkadix	1999	52 bacterial strains with nematocidal activity
CR-USA	1999	One compound with significant anti-malarial activity
LISAN	2000	Two phytopharmaceuticals in the process
Caraito	2000	None yet
Follajes ticos	2000	None yet
Bougainvillea	2001	None yet
La Gavilana	2001	None yet
Agrobiot	2001	None yet
SACRO	2002	None yet

Source: Tamayo et al. Forthcoming in 2004.

TABLE 3. Contribution to Biodiversity Conservation in Costa Rica (US$ x 1,000)

	1993[1]	1994	1995	1996	1997	1998	1999	2000	Total
Ministry of Environment and Energy	110	43	67	51	95	24	39	87	516
Conservation Areas (Development of Bioprospecting Research)	86	203	154	192	126	30	0	0	791
Costa Rican Public Universities	460	126	47	31	35	14	7	4	724
Other groups in INBio	228	93	118	173	129	0	0	0	741
Total	884	465	386	447	385	68	46	91	2,772

[1] Estimated amounts since 1991.

Article 6 (public domain) establishes that the biochemical and genetic properties of the components of wild or domesticated biodiversity belong to the public domain. The State authorizes the exploration, research, bioprospecting, use, and utilization of biodiversity components that are in the public domain, as well as the use of all genetic and biochemical resources, through access standards established in Chapter V of this Law. Likewise, in accordance with Articles 62 and 69, all research or bioprospecting programs on genetic or biochemical biodiversity material to be carried out in Costa Rican territory require an access permit, unless for exceptions provided under this Law. These exceptions (Art 4) basically refer to access to human genetic resources, the exchange of genetic and biochemical resources, and the traditional associated knowledge resulting from the traditional practices of indigenous peoples and local communities when they are non-profit. Public Universities have one year (up to May 7 1999[3]) to establish their own controls and regulations for their research that imply access and are non-profit. If it is not so, all the sectors (pharmaceuticals, agricultural, crop protection, biotechnology, ornamental, herbal, etc.) accessing genetic components are subject to the Law's application and should follow the access procedures.

In this regard, the access regulations are applied to genetic resources on public or private lands, terrestrial or marine environments, under *ex situ* or *in situ* conditions, and in indigenous territories.[4] Likewise, relevant access provisions of the Law are applied to indigenous territories, but additionally their own rules should be taken into account, as well as *sui generis* community intellectual rights. Similarly, communities and indigenous peoples have the right to oppose access to their resources and associated knowledge for cultural, spiritual, economic, or other reasons.

Difficulties and Challenges for Implementing Legal Frameworks: The Case of Costa Rica[5]

In 1998 Costa Rica enacted its Biodiversity Law. It regulates access to genetic and biochemical resources and the sharing of benefits resulting from their use. Costa Rica faced a number of difficulties and challenges in developing its Biodiversity law.

First, there was uncertainty about the scope and value of biodiversity. Bioprospecting is very uncertain; the word "bioprospecting" was derived from prospecting for oil and minerals, but prospecting for biological and genetic resources—or even for indigenous knowledge—is quite different. The risks are greater. Although many samples have been collected from all over the world since the mid-1980s, only a few products have reached the clinical or even pre-clinical stage. When determining the value of genetic resources it must be remembered that the significance of one sample in the overall chain of efforts and costs required to develop a new product or a new drug is very limited. If a country can add value to these resources (e.g., by scientific research), then their value and benefits can increase. Technology has had a paradoxical impact on the value of biological resources. On the one hand, new technologies increase the potential for the commercial use—and thereby the economic value—of biological resources because the cost of screening these materials and/or isolating active ingredients is decreasing. On the other hand, technological developments have reduced the amount of material needed for research purposes, and this may facilitate illegal collection and use. In general, the economic value of genetic resources is increasing, but the commercial value of any particular extract or sample is not.

Second, in regards to property rights and ownership, the CBD does not address the question of ownership; it only establishes (Article 3) that states are sovereign over their genetic and biological resources. But sovereignty, national patrimony, and ownership are different concepts; therefore, it is important to clearly define ownership in the national law. In fact, some of the most common problems that arise when negotiating benefit sharing agreements are related to a lack of clarity about ownership. In Costa Rica, the Law divides biodiversity property rights into genetic and bio-chemical properties and the biological resources per se. Biochemical and genetic properties belong to the State and are therefore under the administration of the Ministry of the Environment and Energy, while biological resources are the property of the landowner, a situation that causes confusion and stirs debates about definitions and use intentions.

Third, a notorious pitfall is over-regulation. The complexity of access regulations creates problems; if nobody can comply with the regulations, then they will likely not be enforced. High transaction costs and bureaucratic procedures also contribute to poor enforcement. Access legislation may negatively affect basic research; it may have negative impacts on local universities and research institutions because basic research is important for conservation purposes and for sustaining biodiversity.

The ultimate goal of access and benefit sharing should be clear. If the main aim is to make money, it is bound to fail. If the objective is to create national capacity, a value added industry, or the conservation of natural biological resources, then it is necessary to make the right connections and develop coherent policies on access, biodiversity conservation, and sustainable use. These policies should include access to knowledge and the traditional use of medical products. Considerations about different treatments or regulations according to the initial nature or purpose of research (non-commercial versus research intended for commercial development) have led to discussions about whether or not to consider all intended research that has the potential to send products into the market place sooner or later.

Lessons Learned

The INBio experience offers a number of lessons related to access and benefits sharing strategies.

First, there must be a clear institutional policy for the criteria demanded in prospecting contract negotiations. For INBio, these include the transfer of technology, royalties, limited quantity and time access, limited exclusiveness, no negative impacts on biodiversity, and direct payment for conservation. This policy has led to the stipulation of minimum requirements for initiating negotiations, and these requirements have meant rejecting some requests (e.g., very low royalties, unwillingness to grant training, etc.). This institutional policy also provides greater transparency and certainty for future negotiations. These same policies must also be taken into consideration when local communities and indigenous peoples, such as the Kuna's in Panama, adopt legal outlines (Cabrera, 1997) in the contractual arrangements entered into by them. They should include other relevant ideas, such as those related to the impossibility of patenting certain elements, licensing instead of a complete transfer, etc.

Second, the existence of national scientific capabilities, and consequently, the possibilities of adding value to biodiversity elements, increases the negotiating strengths and benefit sharing stipulated in contract agreements. As we previously mentioned, the need to grant an aggregated value to material, extracts, etc., is crucial if one wishes to be more than just a simple genetic resource provider. In this regard, the development of important human, technical, and infrastructure capacities through laboratories, equipment, etc., together with the institution's prestige, have permitted better negotiation conditions. The existence of relevant traditional knowledge for operations, which INBio has not yet experienced, implies greater scientific capacity and, consequently, should lead to better compensation conditions.

Third, knowledge of operational norms and of the changes and transformations taking place in the business sector, as well as the scientific and technological innovations that underlie these transformations, helps to define access and benefit sharing mechanisms. It is essential to know how different markets operate and what access and benefit sharing practices already exist in these markets. These vary from sector to sector: the market dynamics for nutraceuticals, ornamental plants, crop protection, cosmetics, and pharmaceuticals are complex and different (see for example Ten Kate and Laird, 1999). This knowledge is needed to correctly negotiate royalties and other payment terms. How can we otherwise know if a percentage is low or high? It is also crucial to be informed about the operational aspects of these markets. When INBio began negotiating new compensation forms, such as advance payments or payments on reaching predefined milestones, with Eli Lilly and Akkaddix, it was vitally important to know the approximate amounts the industry was likely to pay in order to negotiate appropriately. Otherwise, one will likely request terms that are completely off the market or accept terms that are inadequate.

Fourth, it is important to have internal capacity for negotiations, which includes adequate legal and counseling skills about the main aspects of commercial and environmental law. The Institute now recognizes that negotiations involve a scientific aspect (of crucial importance to define key areas of interest such as a product, etc.), a commercial aspect, a negotiation aspect, and the respective legal aspects. These latter are comprised not only of national trade law but also international environment law, conflict resolution, and intellectual property. For these reasons, creating interdisciplinary teams is crucial (Sittenfeld and Lovejoy, 1998). At the same time, the need for such a team is one of the most important criticisms of the contractual mechanisms. Solutions such as facilitators or others that pretend to "level the negotiation power" have been proposed by several authors. Unfortunately, until appropriate multilateral mechanisms exist, benefit sharing and contractual systems must go hand in hand. The absence of an interdisciplinary team keeps one of the parties at a disadvantage, particularly given the enormous legal and negotiation capabilities of pharmaceutical companies.

Fifth, there are a variety of innovative and creative ideas for obtaining compensation. An ample spectrum of potential benefits exists. In the past, interesting benefit sharing formulas were developed through appropriate negotiations. Such formulas included, for example, fees for visiting gene banks, collecting material, etc. The contractual path fortunately permits parties to adapt themselves to the unique situation of each concrete case and to proceed from there to stipulate new clauses and dispositions.

Sixth, it is important to have a deep understanding in such key subjects as: intellectual property rights; the importance of warranties for legality; clauses on ways to estimate benefits (net, gross, etc.); requirements and restrictions on third party transference of material (including subsidiaries, etc.) and the obligations of such parties; precise definitions of key terms that condition and outline other important obligations (products, extracts, material, chemical entity, etc.); precise determination of property and ownership (IPR and others) of the research results, joint relationships, etc.; confidentiality clauses in the agreements and how to balance them in relation to the need for transparency in the agreement; termination of obligations and the definition of the survivor of some obligations and rights (e.g., royalty, confidentiality, etc.); and conflict resolutions. As sub-clause D makes very clear, negotiated agreements are complex. For example, the outcomes that give rise to benefit sharing, such as royalties, will depend on the nature of the definitions for "product", "extract", "entity", etc. A more comprehensive definition will lead to a better position. Further examples of aspects that must be specified include delimiting the areas or sectors where samples can be used, the net sales, and what is possible to exclude from them. In addition, the procedures and rights in the case of joint and individual inventions are of interest (preference and acquisition rights, etc.), as are the conditions for the transfer of material to third parties (under the same terms as the main agreement? need for consent or information? transference to third parties so that certain services can be performed? etc.).

Seventh, it is often critical to adopt a proactive focus according to institutional policies. There is no need to remain inactive while waiting for companies to knock on the door to negotiate. An active approach to negotiations based on the institution's own policy for understanding national and local requirements has produced important benefits. INBio's Business Development Office and its highly qualified expert staff, the attendance at seminars and activities with industry, the distribution or sharing of information and material, and direct contacts, all empower our institution to deal with challenges. The current policy is based on the idea that it is not enough to wait to be contacted or to be available at the behest of a company; instead, one should possess and maintain one's own approach.

Eighth, it is necessary to understand national and local needs in terms of technology, training, and joint research. International strategic alliances must be struck. Even when an institution or community possesses adequate resources to face a concrete demand, knowing the national situation and the strategic needs will permit it to reach better agreements and fulfill a mission that goes beyond merely satisfying the institution's interests. It will permit the prospecting to benefit society as a whole and demonstrate that it is possible to improve quality of life.

Ninth, for prospecting to succeed, so-called macro policies have to exist (Sittenfeld and Lovejoy, 1998); that is to say, there must be clear rules about the "bioprospecting framework," which requires biodiversity inventories, information systems, business development, and technology access. One reason for Costa Rica's success is that institutions not only have experience in negotiation but also in setting policies and actions in this area overall. This includes, for example, a current biodiversity inventory rated as "successful" that enables us to know what we possess. It is the first step in the quest to use this resource intelligently. Our relevant experience also includes a National Conservation Area System that assures the availability of resources, the possibility of future supplies and provisions, mechanisms that contribute to the conservation of biodiversity as part of the contractual systems, and so on. At the same time, the possibility of possessing adequate instruments to manage information, systems of land and property ownership, and so on contribute jointly with the existing scientific capacity to create a favorable environment for bioprospecting and to make possible the negotiation and attraction of joint enterprises. To this should be added other elements, such as the existence of trustworthy partners, which is one of the most relevant aspects in joint undertakings (see Sittenfeld and Lovejoy, 1998).

Lastly, one crucial topic is the constant denouncement of the business community because of the uncertainty caused by the new access rules (mainly in terms of who is the competent authority, the steps to be taken, how to secure prior informed consent, etc.). The emergence of these new regimes, together with the fact that the intention is to essentially control genetic information, its flow, supply, and reception—a topic where little national, regional, and international experience exists—has caused concern because of the possibility of contravening legal provisions. This has led to the establishment, as a policy, of the inclusion of clauses related to the need to fulfill local regulations, to demonstrate the contracting parties' right to fulfill their obligations pursuant to national laws, to present the appropriate permits and licenses, etc. In some cases, this topic has generated important discussions and analyses in negotiations. At an international level, various bio-prospecting agreements around the world are the target of complaints, claims, and lawsuits precisely due to the lack of legal certainty. This has created problems and discrepancies that hinder activities and joint ventures. A few examples would be complaints about the Agreement between Diversa and the Autonomous University of Mexico (which is still being litigated); the deal between this company and Yellowstone National Park; and criticisms of the agreement between the Venezuelan Ministry of the Environment and the Federal University of Zurich.

CONCLUSIONS

The case of Costa Rica has interesting individual features that make it worthy of consideration, but it is not necessarily an example to be followed by other nations. The peculiar circumstances of its national reality (see Mateo, 1996 and 2000 for these special situations), the size of the country, the structure of the central government, and its political, educational, and social situation, among others, have led to the establishment of important but unique conditions. It is, however, an example of a nation that chose a path instead of continuing to discuss the difficulties of potentially traveling on one. From this perspective, the practical experiences of access and benefit sharing embodied in contracts and collaboration treaties with the public and private sectors at the national and international levels, the creation of a Law of Biodiversity that seeks to answer the challenges made by the Convention, the regulation of general *sui generis* systems principles, etc., are all concrete elements to ground further debate. This is probably the most valuable aspect of our experience.

NOTES

[1] The opinions expressed in this paper are the personal opinions of the author and do not necessarily represent those of INBio.

[2] On many occasions conflicts have even arisen because of patents granted to different firms that overlap or because the utilization of a product or process leads to confrontation with different patent holders (e.g., technology used, promoters, etc.)

[3] Only the University of Costa Rica developed its own Regulation of Access regimes.

[4] Article 2 (Area of application) of the Draft Regulations on Access states that it shall be applied on genetic and biochemical elements of wild or domesticated biodiversity, *in situ* or *ex situ*, under State Sovereignty, that are public or private propriety.

[5] This section principally draws from Sittenfeld et al. (2003).

REFERENCES

Cabrera MJ (1997) Contratos Internacionales de Uso de Diversidad Biológica. Una nueva forma de cooperación Norte-Sur, Revista de Relaciones Internacionales 56–57. Escuela de Relaciones Internacionales de la Universidad Nacional, Primer y Segundo Semestre de 1997. Costa Rica: Heredia

Krattiger A (2000) An Overview of ISAAA from 1992 to 2000. *ISAAA Briefs* No 19. ISAAA, Ithaca, N Y

Mateo N (1996) Wild biodiversity: the last frontier? the case of Costa Rica. In: Bonte-Friedheim C et al (eds) The Place of Agricultural Research. ISNAR, The Hague

Mateo N (2000) Bioprospecting and conservation in Costa Rica. In: Svarstad H and Dhillion S (eds) Responding to Bioprospecting. From Biodiversity in the South to Medicines in the North. Spartacus, Oslo

Sittenfeld A, Lovejoy A (1998) Biodiversity prospecting frameworks: the INBio experience in Costa Rica. In: Guruswamy LD, McNeely JA (eds) Protection of global biodiversity: converging strategies. Duke University Press, Durham

Sittenfeld A, Cabrera J, Marielos M (2003) Bioprospecting frameworks: policy issues for Island countries in Insula. International Journal of Island Affairs 12(1)

Tamayo G, Gàmez R, Guevara L (2004) Biodiversity prospecting the INBio Experience. In: Bull A (ed) Microbial Diversity and Bioprospecting, ASM Press, Washington, 445–449

Ten KK, Laird S (1999) The commercial use of biodiversity. Access to Genetic Resources and Benefit-Sharing. Earthscan Publications, London

PART FOUR

ACCESS AND BENEFIT SHARING IN THE NEW MILLENNIUM

CHAPTER 10

CONCLUSIONS: NEW PATHS TO ACCESS AND BENEFIT SHARING

Over the past few years many policy groups and government agencies have made formal recommendations about how the patent system should be reformed (e.g. Nuffield Council, CBAC 2002, Australian Law Reform Commission 2004, UK Government). Many of their recommendations related to Access and Benefits Sharing (ABS) are surprisingly similar. They all argue that there is a need to clarify research and use exemptions (particularly for medical applications), to consider some form of compulsory licensing in areas where patent rights are too restrictive or anticompetitive and to tighten the utility requirements to avoid both biopiracy and the anti-commons.

While none of those specific recommendations has seen action, there has been some modest change in the domestic operation of patent systems in various OECD countries that have at least partly addressed some of the ABS concerns raised. The US, for example, revised its requirements with respect to the utility criterion in January 2001, now requiring patent claims to identify specific uses. The United States Manual of Patent Examining Procedure (MPEP) provides very detailed guidelines concerning the policies and procedures to be followed by staff in the examination of patent applications. These guidelines are public and available for applicants and other stakeholders to better understand how patent criteria are applied. Meanwhile a number of countries are addressing access questions more directly, through adoption of experimental use exemptions. The European Union (EU), for instance, already has a liberal experimental use exemption and the OECD (2003) reports that a variety of other countries are contemplating adopting research or experimental use exemptions within their patent laws.

Others are approaching the ABS issue through more inclusive challenge systems. The TRIPs Agreement permits an *ordre public* provision to incorporate non-economic values into the patent system. Japan and some member states of the EU have adopted related measures. On a case-by-case basis, patents can be refused should the commercial exploitation of the invention violate public order or morality. The European Directive 98/44 on the Legal Protection of Biotechnological Inventions, for one, explicitly states that processes to use human embryos for commercial purposes or processes to clone human beings violate *ordre public* and morality. In practice, the *ordre public* provision is usually invoked by a third party in an opposition procedure after the patent has been granted. Others provide a broader range of opportunities to challenge patents, including an opposition procedure which provides a forum for raising challenges, typically in terms of novelty and inventiveness, but also with respect to *ordre public*. Australia, the European Patent Office, France, Germany, India and Japan currently have opposition processes and a recent US report has recommended establishing an opposition procedure in the US patent system. Even in absence of a more open opposition procedure, the US has established the Court of Appeals for the Federal Circuit to increase uniformity in appeal decisions.

Finally, many governments and agencies are attempting to improve the practice of licensing. The US National Institutes of Health (2004) has developed and the OECD (2005) is in the process of developing guidelines on licensing of human genetic inventions. These guidelines aim at providing a non-binding, but morally persuasive, set of principles and best practices to assist industry and universities in negotiating license arrangements that serve both the interests of industry and the public at large, including the health care sector. Likewise, the World Intellectual Property Organization (WIPO) and the International Trade Centre (ITC) have a new, practical guide on negotiating technology-licensing agreements (WIPO/ITC, 2005).

While there appears to be movement, a complicating factor in this area is that there are now multiple new actors engaged in the debate and discussion. In the past, discussion about the patent system was restricted to a narrow group of technical experts, including patent attorneys, patent examiners, inventors and their assignees. Now citizens, consumers, non-governmental organizations, academics and a wider array of governmental policy actors (both domestically and internationally) are demanding standing in the debate. "Democratization" of patent discussions is leading to more and different questions being asked and raising expectation that policy processes could deliver an international patent system that is more accessible and equitable.

In the context of our analysis, it is vital to acknowledge that many of the problems identified related to ABS are not unique to patents on biotechnology but are common to all areas of new technologies or new markets. New markets are always characterized by uncertainty and high transaction costs, which tends to affect access and more equitable distribution of benefits. Ultimately, our work suggests strongly that many of the often cited problems related to ABS and remedies related to access and benefits sharing are inappropriate. Before assuming change is needed, it is important to consider counterfactual and comparative circumstances. In essence, two specific questions should be asked in any further work around patents. First, the counterfactuals to patents should be considered: "If there were no patents, then...?" Given the interlocking nature of all the IPR regimes, the answer might be "not much." Second, one should consider whether replacing living matter with another type of invention (e.g. nails) would yield the same concern: "Life science patents cause..." could be replaced by "Patents on nails cause..." The nature and scope of the answers is more likely to yield new insights than much of the debate underway today.

RECENT MOVES TO EXTEND ABS

As earlier mentioned, developing countries (often referred to as source countries for genetic resources) have continued to lobby at international institutions for the extension of IPR status to traditional knowledge and plant genetic resources.

The lobby appears to have yielded some results as traditional knowledge is now recognized and included in the agenda of some international institutions and multilateral fora. Some of the key institutions or international agreements that have blazed this trail include World Intellectual Property Organization (WIPO), the Convention on Biological Diversity (CBD), the FAO International Undertaking on Plant Genetic Resources (IUPGR), the recent International Treaty on Plant Genetic Resources (also known as the Global Seed Treaty), and the International Union for the Protection of New Varieties of Plants (UPOV).

Nevertheless, none of the above mentioned institutions or multilateral agreements can elevate traditional knowledge over PGRs to any status that will enable the communities or source states that hold this knowledge to enjoy patent protection on them. Only the WTO has the power and authority to administer the IPR regime instituted through the TRIPs Agreement. Since the TRIPs Agreement is not made subordinate to any other multilateral agreement, the only way that traditional knowledge associated with PGRs can acquire the *toga* of patentability is through amendment of the TRIPs Agreement. This seems a long route to go in view of the interests that will be affected, the conflicts that will arise with a redefinition of the criteria for patents under this circumstance, the question of the commons concept, the debate on the unorthodox mechanism for determining traditional knowledge and the free exchange of knowledge amongst the communities in comparison to the secrecy that drives IPR.

Notwithstanding the above concerns and fears, there are indications that the patent system may be overhauled in the light of paragraph 19 of the Ministerial Declaration issued at the end of the Fourth WTO Ministerial Conference, held in Doha, Qatar, from 9 to 14 November 2001. The Declaration directs the Council for TRIPs responsible for reviewing Article 27.3(b) of TRIPs Agreement (dealing with issues on patenting or otherwise of genetic resources) to "examine *inter alia* the relationship between the TRIPs Agreement and the Convention on Biological Diversity (CBD); the protection of traditional knowledge and folklore; and other relevant new developments raised by members pursuant to Article 71.1 of TRIPs" (Titled: Review and Amendment of TRIPs Agreement).

The Doha Declaration opened up new issues for negotiation by WTO members. Though not specifically mentioned, the principles of access and benefit sharing under the CBD could become the focal point for any harmonization initiative during this round. This is assuming that harmonization of the TRIPs Agreement and the CBD would address rather than exacerbate the current imbalance between holders and tenders of germplasm and the associated traditional knowledge. If in the unlikely event that the ongoing negotiation of the Doha Declaration gives rise to the amendment of the TRIPs Agreement, there is no guarantee that such amendments would elevate traditional knowledge to an intangible patentable item. The Multinational Corporations (MNCs) who use genetic resources and the traditional knowledge about them would surely block such development. In addition, Western states will object to any changes that would automatically alter the fundamental assumptions in IPR definitions of what qualifies for patent protection. This is because to do otherwise would jeopardize the economic foundation for determining patentable subject matters. The criteria remain novelty, inventive step, and the possibility of industrial value. Any revolutionary changes would seriously challenge the contemporary economic system.

As Strange (1994, 24–5) noted, structural power "is the power to shape and determine the structures of the global political economy within which other states, their political institutions, their economic enterprises and (not least) their scientists and other professional people have to operate." The United States and its MNCs enjoy a greater part of the structural power in the world. Since consensus at international institutions is influenced by these structural powers, it is doubtful if developing countries can on their own turn the Doha Agenda into a binding agreement within TRIPs.

Given the above situation, developing countries and those sympathetic to the recognition of plant genetic resources and the associated traditional knowledge as patentable items face an uphill task. Some problems which developing countries must grapple with in the current round of WTO negotiation include lack of organized trade groups or communities with sufficient funds to match the biotechnology lobbying capacity of MNCs and lack of consensus on the international structure and mechanism for benefit sharing. Each source country is currently struggling with implementing its own national schemes and unable to contribute to an international effort. Part of the problem is also the internal conflict on who should exercise proprietary control over PGRs and the associated traditional knowledge. While some argue that sovereignty in such resources resides in the indigenous communities (Shiva, 1991), others, including international conventions (e.g. article 15 of the CBD), acknowledge that the state is the holder of the sovereignty over such resources. Thus, at the negotiation, agreeing on who "owns" the natural resources may be a problem which developing countries may be incapable of resolving; this is complicated because many of the potential properties were developed generations ago, so there are no obvious claimants to work with.

While the process unfolds, it should be noted that the United States could strongly resist any agreement that would diminish its profits or access to raw PGRs for its biotechnology corporations. Lane (1995) reports that when President Bill Clinton signed the CBD in 1993 he assured US biotechnology companies that "his administration would protect intellectual properties according to the standards of the agreement on Trade-Related aspects of Intellectual Property (TRIPs) of the Uruguay Round of the General Agreement on Trade and Tariffs (GATT), and that it would not *consent to a legally binding protocol to the Convention regulating the handling, transfer and use of biotechnology products*" (emphasis added). There is nothing to suggest that this commitment has lessened under the present administration.

Notwithstanding the above huddles, some are convinced that the level of pressure that compelled the WTO to incorporate paragraph 19 into its Ministerial Declaration at the 2001 meeting will also push for its successful negotiation. Admittedly, there are powerful non-governmental organizations and other interest groups backing developing countries in this regard. However, the extent of these organizations' power to influence a major economic change in the international political economy is yet to be seen.

The likelihood of the negotiation succeeding depends on the number of countries that agree to collaborate with developing countries on this issue. Based on prior activities and political leanings (including the colonial affiliation of most developing countries to the EU), the EU offers a greater potential for effective collaboration with developing countries in negotiating for an extension of TRIPs to include traditional knowledge. One indication of this is that the EU is one of the more enthusiastic supporters of the Biosafety Protocol to the CBD, which came into force on September 11, 2003. The EU is one of the few parties to demonstrate appreciation of the "cultural perception of intellectual property" (Drahos, 2001, 426). With 27 present and more potential members who can collectively exercise their votes in the process, the EU might be able to swing a vote in the WTO.

While there are significant differences between the efforts through the Biosafety Protocol and the CBD on the one hand and the TRIPs Agreement on the other, the potential for the EU to help bridge those differences offers hope that this agenda item may see action in the current international negotiations.

ACCESS AND BENEFITS SHARING IN THE NEW INTERNATIONAL IP ORDER

It is against this backdrop that we discuss the practicality of the principle on access to genetic resources and benefit sharing of biotechnology gains enshrined in the UN Convention on Biological Diversity (CBD). We contend that although paragraph 19 of the 2001 WTO Ministerial Declaration provides developing countries an opportunity to negotiate for the expansion of the IPR regime to accommodate community rights, these countries lack the requisite structural power to influence such a drastic change. Developing countries need a new or different international platform to realize the ABS objectives in the CBD.

Generally, the concept of sharing the benefits of any venture, arrangement or collaboration is not new. The manner of making such a distribution depends on the existing agreement between the parties. If it were an incorporated entity, the shareholders' entitlement to any dividends would be shared in proportion to their investment. However, corporations may only declare dividends where there has been a profit, as most jurisdictions prohibit the payment of dividends from capital. Similarly, partners are entitled to partnership drawings but only to the extent of their partnership involvement. Within the public sector, the arrangement is slightly different. Being a utility maximizing umpire, governments are generally held accountable and expected to be fair, just and equitable in their public good distribution responsibilities.

The principle on access to genetic resources and benefit sharing in the CBD is more complicated than the above simplistic explanation of profit sharing in a private enterprise or through a domestic government system. While Article 19(1) of the CBD creates an obligation to adopt policies and measures, including legislation, to ensure source countries' effective participation in biotechnology research and

Article 19(2) requires the promotion of priority access to enable source countries to obtain the results and benefits of biotechnology inventions from any genetic resources provided, it has been difficult to translate these lofty goals into practical benefits. Generally, source states have faced three challenges in trying to enforce their ABS rights. First, none of the key terms in the ABS area is defined in the CBD. Second, there is still a persistent notion in the West that genetic resources found in developing countries are part of the common heritage of humankind and therefore should be freely accessible to all humankind, without discrimination (incidentally, the United States interprets such undisturbed access as benefit sharing). Third, Article 15(7) of the CBD stipulates that benefit sharing should be at "mutually agreed terms"; this bilateral approach ignores the power imbalance between source states where the genetic resources are found and the biotechnology countries that utilize these resources. Some argue that an international regime would be more effective. In a few countries where genetic resource users are willing to negotiate with the source states, the latter have often found it difficult to strike a sustainable bargain. To date, the ABS model in Costa Rica is one of the few effective benefit sharing arrangements within the CBD structure.

With respect to the key terms, none of "access", "benefit" or "sharing" is defined in the Convention. This is one occasion when it would have been most useful to clarify terms that drive the ABS principle. Several questions are left unanswered. These include what does "access" under the Convention comprise of? What level of access is acceptable under the CBD? Is partial access adequate? What would be full access? The Convention also does not provide any criteria for measuring success on the workability of the principle on access to biological resources. Consequently, nothing stops an entity granted access to re-evaluate the relationship and decide that it has not received substantial access to the resources. If this happens, such an entity would have the basis for refusing to share the benefits attached to the purported limited access obtained. In the same vein, the source state may be unable to negotiate substantial benefits for itself as, from the beginning, it lacks the technical know-how to adequately quantify the value of the resources which it is allowing access to. Incidentally, there is no mechanism for measuring such value for some types of biological resources and the associated traditional knowledge.

Similarly, it is unclear if monetary or non-monetary compensation would qualify as "benefit" under the Convention. Assuming non-monetary compensation is acceptable, what should be the nature of this in-kind payment? Does allowing the source state access to the technology used in the process suffice? What of scholarships to indigenous people of the source states studying in relevant fields? How do you determine the beneficiaries of such scholarships? If technology transfer is an acceptable benefit, what should be the measure of such technology be in order to avoid the transfer of obsolete technology? Lastly, would a joint venture arrangement between source states and users of their genetic resources and associated traditional knowledge be sufficient for this purpose? It is possible all the above possibilities and more could suffice. The lack of description of what an appropriate "benefit" would be has confined this issue to the bargaining table, where little has been accomplished.

The word "sharing" suggests a division, partition or distribution of some sort, which must relate to advantages, profits or benefits obtained from the utilization of biological resources. The Convention requires the distribution to be on a mutually agreed terms. Since "access" to the genetic resources would precede the "sharing" of the gains from biotechnology inventions arising from their uses, it is not clear what happens when the genetic resource receiving party refuses to share the benefits. CBD is silent on this point and, most likely, such matters would be resolved under the contract between the parties or by *force majure*.

There are some structures that offer a starting point for developing an operational terminology. The Common Policy Guidelines for participating Botanic Gardens drafted in 2000 defines "access to genetic resources" as "the permission to acquire and use genetic resources." It also defines "benefit-sharing" as "the sharing of benefits arising from the use, whether commercial or not, of genetic resources and their derivatives, and may include both commercial and non-commercial returns." Access to biological resources may only be possible with the consent of the source state in which the biological resources are found. This means that the resources must be found within the territorial boundaries of a source state before its permission can be sought. Although it could be implied, the Guidelines' definition does not deal with the fundamental problems of joint ownership of biological resources where one country may give consent and another withholds consent. There is no clue on how to resolve such impasse. It is even more problematic determining which international agency should handle an issue of this nature as the matter impinges on the sovereign rights of both source states.

The Guidelines' definition further ignores one of the main controversies surrounding the concept of benefit sharing—that is the persistent notion in the West that biological resources found in developing countries are part of the commons over which no country should exercise specific ownership. The commons debate, which is the focus of Chapter 2, is a very real obstacle to the realization of the ABS objectives. Some argue that it is inequitable to demand full sharing of common heritage resources, the use of which does not diminish or extinguish other peoples' right to those resources. Others contend that since the use to which the resources may be subjected would benefit the whole of humankind, the corporations involved in the R&D should be compensated not penalized. There are also those who are of the view that it should be sufficient for users of these resources to disclose their sources of origin. Such disclosures should be made at the time of application for IPR protection. Neither the Bonn Guidelines on access to genetic resources and fair and equitable sharing of the benefits arising from their utilization nor the resolution reached at the 2002 World Summit on Sustainable Development held in Johannesburg has assisted source states in achieving the CBD objectives.

There is still an ongoing discussion on the best way to implement the benefit sharing provision of Article 8(j). Although the bulk of the responsibility is on states to fashion what access and benefit-sharing arrangement suits their national objectives, no workable solution is yet in place. Currently, national structures on access and benefit sharing are mainly driven on a case-by-case basis. There has been little success within the broad ideal of ABS instituted under the CBD. This explains why the CBD's Ad Hoc Open-Ended Working Group on Access and Benefit Sharing (the "Working Group") continues to work tirelessly for solutions. It is therefore not surprising that the Working Group resolved in their December 2003 meeting to recommend to the Conference of the Parties (COP) for permission to enter into negotiations on an international regime on both access and benefit sharing. The parties approved this recommendation at their COP-7 meeting in February 2004 and set the terms of reference on which the negotiations will be based.

For the most part, the access part of the ABS scheme seems not to be a problem as genetic resource users find ways to circumvent the CBD threshold. This is why accusations of biopiracy are still rampant. Some developing countries regard the Doha Development Agenda as the conduit for achieving the CBD objective. But this can only be the case if the agenda is successfully negotiated to cater for such goals. The harmonization envisaged under the Doha Development Agenda between the CBD and the TRIPs Agreement is unlikely to happen if the result would hamper private property rights. The broad terms of the CBD on ABS will enjoy the United States' support only so long as they do not impact IPRs. After all, the United States is yet to ratify the CBD. Thus, even though patentability of traditional knowledge is one of the issues that the TRIPs council will review at the ongoing Doha driven multilateral trade negotiations, it is possible that the task of designing the workability of the ABS scheme will be passed to other UN agencies such as the WIPO or the CBD Working Group on Article 8(j). At its Twenty-sixth (12th Extraordinary) Session held in Geneva between September 25 and October 3, 2000, WIPO's secretariat prepared a document that clearly argues for recognition of traditional knowledge associated with "biotechnology and genetic resources." The document is modeled after the CBD with details on equitable sharing of gains and benefits from gene-technology.

Suffice to say that any changes to the current IPR regime under the TRIPs Agreement will require a powerful lobby of the nature used by the United States and its business community during the Uruguay Round. It is doubtful if the developing countries have such organizational ability to fully negotiate the expansion of the existing patent regime to include traditional knowledge. These countries do not have substantial structural power (defined by Strange 1994 as the power to define the rules that determine outcomes) to organize, nurture and sustain such a range of alliances. It is even more difficult given the duration that it takes to fully negotiate an issue at the multilateral institution. Lastly, these countries are incapable of raising the required funds to undertake such exercise.

CONCLUSION

Although there is no scientific method for determining the likely outcome of the ongoing negotiations of paragraph 19 of the Doha Declaration, recent events, including the inconclusiveness of the Cancun Ministerial, suggests that not much will be achieved. Each group—developed and developing countries—continues to approach the negotiation with keen interest and careful preparation. Although developing countries had nearly 5 years to prepare for this round of WTO negotiations, there is no indication that they can present a united front at the negotiating table. They are largely divided on the scope and nature of the changes they seek to bargain. While some of them would like to see a *"sui generis"* system that is tailor-made for their specific needs" within the TRIPs regime, others support an alternative approach that will be independent of that agreement. Whatever, they finally settle for, "it remains to be seen to what extent developing countries can formulate a strategy to pursue their interests within the TRIPs Agreement" (Anon, 1998)

Most likely, regional affiliations such as the North American Free Trade Association (NAFTA) could play a big role in defining how countries side on different issues (Hveem, 2000). It is important to keep in mind that any new bargain will not be just amongst equal nation states. Rather, it will be a complex balance between weak states and their partners on the one hand and powerful MNCs and a handful of wealthy states on the other. At the moment and for the foreseeable future, the balance of power is largely in favour of the latter. It is likely that those with structural power will use their power to project and safeguard their interests. It would not be unusual for the US and other OECD states to apply a "carrot and stick" approach similar to what happened during the Uruguay Round. Generally, the US does not hesitate to apply measures if the interests of its corporate citizens are threatened by changes in either a legal order or institutions. As Gilpin (1987, 241) notes:

> Although the interests of American corporations and U.S. foreign policy objectives have collided on many occasions, a complimentarity of interests has tended to exist between the corporations and the U.S. government. American corporate and political leaders have in general believed that the foreign expansion of American corporations *serves important national interest of the United States*. American policies have encouraged corporate expansion abroad and have tended to protect them.

At stake in this negotiation are the ownership, control and enjoyment of raw plant genetic resources and associated traditional knowledge. Presently, genetic resources are erroneously described as the common heritage of humankind, and therefore not capable of private ownership. In contrast, the products that result from scientific processing of these resources and knowledge enjoy patent protection. To redefine the patent regime will require changes in institutions and regulations and a diminution of the knowledge power controlled by MNCs.

The Uruguay Round engaged a new set of powerful private-sector stakeholders. They were highly influential in the past round, effectively crafting the TRIPs Agreement, and there is no evidence that they will be any less active and powerful in the current negotiations. Changes in technology and markets have continued to favour these stakeholders and they are likely to vociferously oppose any changes that they would view as inimical to their prosperity. These actors, and their home states, are seldom driven by sympathy. Developing countries will need to offer WTO members something beyond rhetoric to gain in the current round. The challenge will be to develop an economic, institutional and commercial basis for ABS that can both unite the donor countries and provide a compelling reason for both MNCs and the developed countries to engage.

REFERENCES

Anon, 1998. Editorial : Trips and the legal protection of plants. *Biotechnology and Development Monitor,* 34, p.2-3. Available at http://www.biotech-monitor.nl/new/index.php?/ink=Publications

Australian Law Reform Commission (2004) Genes and ingenuity: gene patenting and human health Biotech Monitor, 1998 ALRC, Canberra. Available at http://www.austlii.edu.ak/au/other/alrc/Publications/reports/99

Canadian Biotechnology Advisory Committee (CBAC) (2002) Improving the regulation of genetically modified foods and other novel foods in canada: a report to the government of canada biotechnology ministerial coordinating committee CBAC, ottawa. Available at http://cbac-cccb.ca/epic/internet//incbac-cccb.nsf/vwapj/Improving-Regulation-GMFoodAug02.pdf/$FILE/Improving-Regulation-GMFoodAug02.pdf.

European Commission (1998) Directive 98/44 of the European parliament and of the council of 6 July 1998 in the legal protection of biotechnological Inventions, OJL (1998) No L213

Hveem H (2000) Explaining regional phenomenon in the era of globalization In: Stubbs R, Underhill GRD (eds) Political economy and the changing global order. Oxford University Press, Oxford

National Institutes of Health (2004, November 19) Best practices for the licensing of genomic inventions. NIH Washington, (for comment by January 18 2005), also available at http://ott.od.nih.gov/NewPages/LicGenInv.pdf

Organization for Economic Cooperation and Development (2005, February 1) Draft guidelines for the licensing of genetic invention. OECD, Paris, also available at www.oecd.org/document/26/0,2340,en_2649_37437_ 34317658_1_1_1_37437,00.html

Organization for Economic Co-operation and Development (2003) Genetic inventions, IPRs and licensing practices: Evidence and policies. OECD, Paris

Shiva V (1991) Biodiversity, biotechnology and profits. In: Shiva V et al (eds) Biodiversity: social and ecological perspectives. Zed Books, London, NJ

World Intellectual Property Organization and International Trade Centre (2005) Exchanging value – negotiating technology licensing agreements, WIP/UPD/2005/237

INDEX

Aarhus Convention on Access to Information, Public Participation and Access to Justice in Environmental Matters, 92
Africa, 26, 31, 41, 117–120, 121, 124
Agenda 21, 32, 55, 87
Airspace, 22
Akkadix Corporation, 185, 187
Americas, 95, 96, 117, 118–119
Andean countries, 29
Anns v. Merton, 149
Antarctica, 29
Ardais Inc., 168, 169–170
Argentina, 52, 61, 186

benefits sharing, 3–16, 65–77, 157–175, 199–206
Berne Convention, 11
Bioassays, 15, 185, 186
Biocolonization, 21, 159
Biopiracy, 16, 68, 71, 72–73, 75, 77, 111–135, 159, 199
Bolivia, 29
Bonn Guidelines, 68, 70, 161, 173–176, 205

CAMBIA, 52
Canada's Plant Breeders Rights Act, 50, 53, 57
Canadian Biotechnology Advisory Committee, 59
Canola, 49, 53, 54, 56–59, 147–148
CARTaGENE, 166
CGIAR, 55, 128, 129
Chagas Project, 186
Charter of Economic Rights and Duties, 39
CIMMYT, 127, 129
classification of knowledge, 9, 32–37, 65, 123
coffee, 117, 121, 124
Colonialism, 114–118, 119, 124–126
Colonization, 83, 115, 118–119, 138
commercial strategies, 6, 33, 49
common heritage of humankind, 15, 25, 27–28, 29–30, 34, 35, 36, 55, 204, 207
Common Policy Guidelines for participating Botanic Gardens, 205

common resources, 22, 38, 44
compensatory justice, 15, 65–66, 68, 69–71, 73, 74, 77
Convention on Biological Diversity (CBD), 3, 21, 31, 55, 68, 71, 85, 87, 89, 98–99, 113, 131, 135, 160, 174–175, 201, 203
Convention to Safeguard Intangible Cultural Heritage, 98
Copyright, 6, 38, 40, 51, 112, 134
Costa Rica, 16, 67, 76, 109, 183–191, 194, 204
Costa Rica's Law of Biodiversity No. 7788, 188
Crucible Group, 41, 137
Customary Law, 27, 82, 84, 88, 91, 94, 105–106, 135

Declaration of San Jose, 86
Declaration on the Right to Development, 87
Decode, 162–163
Diamond v. Chakrabarty, 6, 52
Diversa, 185, 194
DNA Sciences Inc., 168–169
Doha Declaration, 201–202
Draft Covenant on Environment and Development (IUCN, 2000), 92
Draft Universal Declaration on the Rights of Indigenous Peoples, 91
duty of care, 148, 149, 150
duty to inform, 145–146, 147, 148, 149, 150–151, 153
duty to warn, 149–150

economic impact of biotechnology, 7
economics of patents, 4–8
Eli Lilly, 185, 192
erythroxylon coca, 123
Estonia, 163–164, 166
Estonian Genome Foundation, 163–164
ethnic communities, 82–83
European, 5, 51, 54, 112, 115–121, 124, 127, 163, 199

INDEX

FAO International Undertaking on Plant Genetic Resources, 201
Food and Agriculture Organization, 28, 55, 70, 127, 131, 183
Forest Principles, 87–88
Framework Convention on Climate Change, 87–88

gene banks, 15, 24, 34, 39, 126–130, 163–164, 193
Genomic Research in the African Diaspora (GRAD Study), 166
Genomics, 65, 159, 166, 169, 175
Gilpin, 9, 17, 207
global commons, 21, 29
GURTS, 56, 60

High Commissioner of the UN Commission on Human Rights, 75
Hoffman and Beaudoin v. Monsanto Canada Inc. and Bayer Cropscience Canada Holding Inc., 16
Human Genome Organization (HUGO), 70–71, 172–173
Human Genome Project (HGP), 157–158, 159, 161, 175
Hybrids, 5, 7, 50, 51, 56, 60, 124

ICARDA, 127
Iceland, 162–163, 164–165
ICRISAT, 127
IITA, 127
ILO Convention 169, 83, 130–131
INBio, 67, 183–186, 188, 191–193
Indigenous Peoples, 82–91, 95–103, 105–106, 111–126, 133–136, 189–190, 191
Intellectual Property Committee, 10, 12
International Convention to Combat Desertification, 103–105
International Covenant on Ecomomic, Social and Cultural Rights (ICESCR), 75, 88, 89
International Declaration on Human Genetic Data (UNESCO, 2003), 98, 173
International Labour Organization (ILO) Convention 169 'Concerning Indigenous Peoples in Independent Countries', 90–91, 92
International Service for the Acquisition of Agri-biotechnologies (ISAAA), 183
International Treaty on Plant Genetic Resources, 55, 201

International Undertaking on Plant Genetic Resources, 55, 201
International Union for the Protection of New Varieties of Plants, 5, 35, 201
Invention, 3–8, 12, 22, 38–39, 51–52, 84, 199, 200
IRRI, 127, 129

Keidanren, 10

Liability, 145–154
Locke, John, 23, 24

Madley v. University of North Carolina, 7–8
medicinal plants, 29, 84, 123, 186
Meeting of Experts on Human Rights and the Environment (2002), 92
Merck and Co., 184, 187, 189
Mexico, 43, 127, 132, 186, 194
Monsanto Canada v. Schmeiser, 15, 49
Myriad Genetics, 146

Nanotechnology, 8
natural right, 15, 37, 73
Nicaragua, 96, 129
Nigeria, 27, 127
Non Governmental Organizations, 86, 125, 200, 202

Organization of American States (OAS), 96
outer space, 26, 29

papavar somniferum, 123
patents for living matter, 3
Peru, 29, 76, 127
Pharmaceuticals, 8, 134, 168, 190, 192
Philippines, 87, 124, 127
plant breeders' rights, 5–8, 15, 35, 49–54, 57, 61, 132, 183
plant genetic resources (PGRs), 9–10, 12–13, 14–16, 21–22, 27–35, 39–43, 131–133, 201, 202, 207
Plant Patent Act 1952, 35
Population Genetics, 157–175
Porter, 9
Potatoes, 121, 124, 127, 128
product liability, 145, 148, 149, 150, 152, 153
property, 158–159, 161, 176, 188, 190–193, 206

Quebec, 166

Rio Declaration, 32, 87
Roundup™, 56

Saskatchewan Court of Queen's Bench, 148
serpent tree, 123
sovereignty of nations, 22
Standpoint epistemology, 69
StarLink, 147, 153
Strange, 9, 11, 202, 206
structural power, 8–11, 43, 202, 203, 206, 207
sui generis system, 52, 93–94, 101, 106, 195, 207
Supreme Court of Canada, 50, 58, 152

terra nullius, 119
tobacco, 117, 123
tort law, 153
trade secrets, 6, 7, 51, 57
trademarks, 6, 7, 38, 40, 50, 54, 56, 57
Traditional Ecological Knowledge (TEK), 32, 37–38, 81–83
traditional knowledge (TK), 15, 32–43, 65–77, 98–99, 101–105, 111, 113, 114, 116, 120, 123–124, 131, 133–136, 200–203, 206, 207
Traditional Resource Rights (TRR), 93–95, 99
Tragedy, 25
Tragedy of the commons, 25
triple helix, 9
TRIPS Agreement, 3, 6, 10–13, 14, 22, 35, 75, 134, 159, 171, 199, 201, 203, 206, 207
TUA agreement, 58

U.S. Plant Variety Protection Act, 35, 51
UK BioBank, 164–166
UNICE, 10
United Nations Conference on the Human Environment, 86
United Nations Environment Programme, 183
United Nations Permanent Forum on Indigenous Peoples, 96–97
Universal Declaration of Human Rights, 75, 88, 89, 126, 171
University of Guelph, 186
University of Massachusetts, 186
UPOV, 1991, 5, 35, 51, 52, 53–56
Uruguay, 61, 186
US Patent Act, 5, 51
USPTO, 73

Vandana Shiva, 10, 35, 68

WARDA, 127
World Conservation Union (IUCN), 81, 92, 98, 99–100
World Cultural and Natural Heritage Convention, 72
World Intellectual Property Organization, 6, 36, 66, 135, 200, 201
World Summit on Sustainable Development, 88, 205
World Trade Organization, 10, 43, 171
WTO Doha Ministerial Declaration, 14

The International Library of Environmental, Agricultural and Food Ethics

1. H. Maat: *Science Cultivating Practice*. A History of Agricultural Science in the Netherlands and its Colonies, 1863-1986. 2002   ISBN 1-4020-0113-4
2. M.K. Deblonde: *Economics as a Political Muse*. Philosophical Reflections on the Relevance of Economics for Ecological Policy. 2002 ISBN 1-4020-0165-7
3. J. Keulartz, M. Korthals, M. Schermer, T.E. Swierstra (eds.): *Pragmatist Ethics for a Technological Culture*. 2003   ISBN 1-4020-0987-9
4. N.P. Guehlstorf: *The Political Theories of Risk Analysis*. 2004
ISBN 1-4020-2881-4
5. M. Korthals: *Before Dinner*. Philosophy and Ethics of Food. 2004
ISBN 1-4020-2992-6
6. J. Bingen, L. Busch (eds.): *Agricultural Standards*. The Shape of the Global Food and Fiber System. 2005   ISBN 1-4020-3983-2
7. C. Coff: *The Taste of Ethics: An Ethic of Food Consumption*. 2006
ISBN 1-4020-4553-0
8. C.J. Preston and Wayne Ouderkirk (eds.): *Nature, Value, Duty: Life on Earth with Holmes Rolston III*. 2007   ISBN 1-4020-4877-7
9. Dirk Willem Postma: *Why Care for Nature? In search of an ethical framework for environmental responsibility and education*. 2006
ISBN 978-1-4020-5002-2
10. P.B. Thompson: *Food Biotechnology in Ethical Perspective*. 2007
ISBN 1-4020-5790-3
11. P.W.B. Phillips, C.B. Onwuekwe (eds.): *Accessing and Sharing the Benefits of the Genomics Revolution*. 2007   ISBN 978-1-4020-5821-9

springer.com

Printed in the United States
99967LV00002B/331/A